Gorges Edmond Howard

A Treatise of the Exchequer and Revenue of Ireland

Vol. 1

Gorges Edmond Howard

A Treatise of the Exchequer and Revenue of Ireland
Vol. 1

ISBN/EAN: 9783337325152

Printed in Europe, USA, Canada, Australia, Japan

Cover: Foto ©berggeist007 / pixelio.de

More available books at **www.hansebooks.com**

A
TREATISE
OF THE
Exchequer and Revenue
OF
IRELAND.

a

Ā

TREATISE

OF THE

Exchequer and Revenue

OF

IRELAND.

By G. E. HOWARD, Esq.

MOST HUMBLY INSCRIBED,

To the TREASURER, CHANCELLOR, LORD CHIEF BARON, and the Reſt of the BARONS of the COURT of EXCHEQUER.

IN TWO VOLUMES.

VOL. I.

DUBLIN:

PRINTED BY J. A. HUSBAND, FOR E. LYNCH, No. 6, SKINNER-ROW.

M,DCC,LXXVI.

SUBSCRIBERS NAMES.

HIS Excellency Earl Harcourt, Lord Lieutenant of Ireland, 6 Sets.

Rt. hon. James Lord Lifford, Lord Chancellor of Ireland.

Rt. hon. William Gerard Hamilton, Chancellor of the Exchequer.

Rt. hon. John Lord Annaly, Lord Chief Juftice of the King's bench.

Rt. hon. Richard Rigby, Mafter of the Rolls.

Rt. hon. Marcus Patterfon, Lord Chief Juftice of the Common pleas.

Rt. hon. Anthony Fofter, Lord Chief Baron of the Exchequer.

Hon. Mr. Juftice Robinfon.

Hon. Mr. Baron Scott.

Hon. Mr. Juftice Tennifon.

Hon. Mr. Juftice Henn.

Hon. Mr. Baron Power.

Hon. Mr. Juftice Lill.

Late hon. Mr. Juftice Malone.

A.

Adams, Mr. James, Commander of the Townfhend Revenue Cruizer.

Agar, James, Efq; one of the Commiffioners of the Revenue.

Annefley, Hon. Richard.

Armfteed, Francis, Efq;

B.

Bellamont, Earl of.

Barry, Gaynor, Efq;

Betagh, Mr. Henry, Attorney.

Beresford, Rt. hon. John, one of the Commiffioners of the Revenue of Ireland.

Birch, Robert, Efq;

Blakeney, Mr. Charles, Attorney.

Blaquiere, Rt. hon. Sir John, 2 Sets.

Bourke, John, Efq; one of the Commiffioners of the Revenue of Ireland.

Browne, Hon. James.

Burgh, Walter, Efq;

Burke, Edmund, Efq;

Burke, William, Efq;

Bufteed, Jephfon, Efq;

Butler, Hon. John.

C.

Charlemont, Earl of.

Cloyne, Bifhop of.

Caldbeck,

SUBSCRIBERS NAMES.

Caldbeck, William, Efq;
Caldwell, Charles, Efq;
Carleton, Hugh, Efq;
Carmichael, Hugh, Efq;
Carroll, Mr. John, Attorney.
Carr, Mr. Richard Cooban.
Carfon, Mr. Robert, Attorney.
Chapman, Benjamin, Efq;
Chefter, Mr. Richard, Attorney.
Clements, Rt. hon. Nathaniel.
Colles, Mr. Richard, Attorney.
Commiffioners of the Impreft Accounts.
Commiffioners of the Revenue, Rt. hon. and hon.
Connor, Charles, Efq;
Cooke, Mr. Theodore.
Copinger, Maurice, Efq; Second Serjeant.
Coulfon, Henry, Efq;
Crawfurd, Gibbs, Efq; Solicitor to the Stamps in England.
Crookfhank, Alexander, Efq;
Crowe, Mr. James, Attorney.

D.
Defart, Rt. hon. Lord.
Dartrey, Lord Baron of.
Damer, John, Efq;
Darby, Jonathan, jun. Efq;
Davis, Jofhua, Efq;
Dennis, James, Efq; Prime Serjeant.
Dobbyn, Robert, Efq;
Dougherty, Mr. John, Attorney.
Dunkin, William, Efq;

E.
Ellis, Rt. hon. Welbore, 2 Sets.

F.
Ferrall, Mr. James, Attorney.
Finucane, Bryan, Efq;

Finucane, Matthew, Efq;
Fitzgerald, Robert, Efq;
Fitzgibbon, John, Efq;
Flood, Frederick, Efq;
Flood, Warden, Efq;
Fofter, John, Efq;
Franklin, Alexander, Efq;
Franks, Mr. Thomas, Attorney.
Frazer, James, Efq;
French, Robert, Efq;

G.
Glafcock, Mr. William, Attorney.
Godley, John, Efq;
Gordon, Robert, Efq;
Gorges, Hamilton, Efq;
Green, Godfrey, Efq;

H.
Hamilton, George, Efq;
Hamilton, Sackville, Efq;
Hart, George, Efq;
Hellen, Robert, Efq; Council to the Commiffioners of the Revenue.
Herbert, John, Efq;
Hewitt, Hon. Jofeph.
Holt, William, Efq;
Hopkins, Francis, Efq;
Howard, Rt. hon. Ralph.
Howard, Hugh, Efq;
Huband, Jofeph, Efq;
Hughes, Francis Annefley, Efq;
Hunter, Charles Orby, Efq;
Huffey, Dudley, Efq;
Hutchinfon, Rt. hon. John Hely, Efq; Provoft of Trinity college.

J.
Jackfon, Mr. William, Attorney.

K.
Keliher, Mr. William, Attorney.
Kelly,

SUBSCRIBERS NAMES.

Kelly, Thomas, Efq;
Kingfbury, Thomas, Efq;
Kirwan, Richard, Efq;

L.
Lane, William, Efq;
Langrifhe, Hercules, Efq; one of the Commiffioners of the Revenue of Ireland.
Lees, John, Efq;
Lennon, Remigius, Efq;
Levinge, Richard, Efq;
Lukey, Mr. George, Attorney.
Lyfter, William, Efq;

M.
Mountmorres, Lord Baron of.
Macarthy, Dalton, Efq;
Macartney, Rt. hon. Sir George, K. B. 6 Sets.
Macartney, Mr. George, Attorney.
Mc. Mollen, John, Efq;
Mc. Nemara, Daniel, Efq;
Malone, Rt. hon. Anthony.
Malone, Richard, Efq;
Mafon, John Monck, Efq; one of the Commiffioners of the Revenue of Ireland.
Maunfell, Thomas, Efq; Council to the Commiffioners of the Revenue.
Montgomery, Vaun, Efq;
Morgan, Richard, Efq;
Morrifon, John, Efq;
Murphy, David, Efq;
Mufgrave, Richard, Efq;

N.
Nafh, Andrew, Efq;
Norton, Brett, Efq;

O.
O'Brien, Sir Lucius, Bart.
O'Brien, Mr. Dennis.
O'Connor, Charles, Efq;

O'Connor, John, Efq;
Ofborne, Rt. hon. Sir William.

P.
Paul, Robert, Efq;
Plumptre, Francis, Efq;
Plumptre, Polidore, Efq;
Ponfonby, Rt. hon. John.

R.
Ratcliffe, Stephen, Efq;
Reilly, Hugh, Efq;
Ridge, John, Efq;
Rowley, Clotworthy, Efq;
Ryan, Matthew, Efq;

S.
Shelburne, Earl of, 2 Sets.
Southwell, Lord Baron.
Scott, John, Efq; Solicitor general.
Sherlocke, John, Efq;
Shiel, James, Efq;
Simpfon, Mr. Richard, Attorney.
Southwell, Hon. Thomas-Arthur.
Smith, Ambrofe, Efq;
Spring, Thomas, Efq;
Stacpole, Jofeph, Efq;
Staples, John, Efq; one of the Commiffioners of the Revenue.
Steele, Sir Richard, Bart.
Sterling, Edward, Efq;
Stewart, Henry, Efq;
Stuart, Hamilton, Efq;
Swan, Edward Bellingham, Efq;.
Swan, John, Efq;
Sweeny, Edward, Efq;
Swift, Mr. Michael, Attorney.

T.
Tighe, Edward, Efq;
Tifdall, Rt. hon. Philip, Attorney general and Secretary of State.
Toler, John, Efq;

Townfend,

SUBSCRIBERS NAMES.

Townfend, Richard, Efq; one of the Commiflioners of the Revenue.

Trant, Dominick, Efq;

Tunnadine, John, Efq;

U and V.

Vernon, George, Efq;

Underwood, Richard, Efq;

W.

Waller, Richard, Efq; Solicitor to the Stamps in Ireland.

Waller, Robert, Efq;

Wallis, John, Efq;

Walfhe, David, Efq;

Wefton, Robert, Efq;

Whittingham, William, Efq;

Williams, Adam, Efq;

Willmott, Sir Robert, Bart.

Wolfe, Arthur, Efq;

Wolfe, Theobald, Efq;

Wood, Attiwell, Efq;

Y.

Yelverton, Barry, Efq;

PREFACE.

PREFACE.

MY only intention, at firſt, in collecting and compoſing the following work was for my own private inſtruction and uſe; for in the year 1743, having been appointed *attorney for the King's rents in the Exchequer*, as I was in the courſe of ſix or ſeven years more to others of thoſe offices which I ſtill hold in the ſeveral legal departments of the revenue *, I very ſoon perceived that the due and proper execution of them required an accurate knowledge of ſeveral matters of which I then was totally ignorant, and with which but very few were acquainted; as the *acts of ſettlement* and *explanation* on the *rebellion* in

---

* And have been lately appointed Solicitor to the commiſſioners of the revenue.

this kingdom in 1641, and the proceedings thereon; the *truflee act* and the feveral after *acts* on the fubfequent *rebellion* in 1688, and the proceedings alfo thereon; feveral branches of the revenue of this kingdom, with the inftitution and conftitution thereof, and the changes therein; as alfo of the feveral offices where the feveral records and archieves relating thereto were to be found.

And being extremely uneafy in this ftate of ignorance of the feveral bufineffes in which it was fit I fhould be knowing, I immediately fet about examining all the books and records in the feveral public offices of the kingdom for upwards of one hundred years before, which were in any fort converfant with the Exchequer and Revenue of Ireland; and fome of the gentlemen who had been for a confiderable time deputies in the faid offices, having not only freely communicated to me every intelligence I requefted of them, but alfo furnifhed me with copies of all fuch extracts, minutes, &c. as they had themfelves taken, or were in their poffeffion, relating to the matters which they found I was collecting. What with thefe, and the innumerable copies and extracts which I had myfelf taken, together with the

many

many fpecial cafes which happened in the court of Exchequer here, relating to my feveral departments in the revenue, in the long courfe of three and thirty years experience, and which I likewife had from time to time collected ; as alfo many fpecial cafes from Englifh authorities ; the whole was at length fwelled to no lefs a bulk in the manufcript, than four very large folio volumes *.

But as feveral arduous matters, not only in the common law, but alfo of the conftitution, (as I may fay) of this kingdom, far above my readings and knowledge, were difperfed through the whole, as I had collected them from various treatifes on

---

* This enquiry brought to my recollection an irreparable lofs to the public, of which I myfelf had knowledge : Dudley Loftus, efq; who had exercifed fome high offices in this kingdom for many years during the reign of King Charles Ift, and the fucceffive Kings, Charles IId, and James IId, a gentleman of great abilities and learning, having made collections of feveral important matters relative to the aforefaid *acts of fettlement* and *explanation*, as alfo to the rebellion in the year 1688, in feveral manufcript volumes, they fell into the hands of my mother, who was a defcendant from him, and fhe having married a gentleman in the army, and they not knowing the value of them, the whole, except one volume, was ufed as wafte paper upon all occafions in the houfe ; however, in this one only volume which happened to efcape, there were a few curious matters relative to the aforefaid *acts of fettlement* and *explanation*, which are inferted in the following work.

A 2   thefe

thefe fubjects, it never was my intention to pre-
fume to commit them to the view of the public,
until feveral gentlemen, not only of eminence,
but of high ftation in the law, having at different
times, within the period of time I have mentioned,
not only in part perufed them, but got feveral
extracts from them, preffed me to do fo.

This, at my time of life, but chiefly for the
other reafons I have mentioned, I could not
think of myfelf to attempt, but thereupon immediately offered to let the public have the work,
if any gentleman of the bar, of fufficient ability,
would give his time and labour to the examination of fo voluminous a collection, and the innumerable authorities from which a confiderable
part of it had been collected, and in reducing
the huge and indigefted mafs to fuch a degree of
method and order as might render it fit for the
eye of the publick; and accordingly for that purpofe, I depofited the four manufcript volumes at
the book ftall in the hall of the four courts,
where they remained for about nine months for
the infpection of all fuch gentlemen of the profeffion, and others, as were inclined to perufe
them.

In

PREFACE.

In fome time after, counfellor Charles O'Neill, whofe learning, knowledge, and abilities, in his profeffion are fo univerfally known, that to expatiate on them here would be entirely needlefs, (after offers had been made by others who were fpeedily deterred from the attempt by the labour and difficulty which it was conceived would attend it) of his own accord, moft kindly undertook to fit the work for the prefs; and although it has greatly interfered with the bufinefs which his merit has fo juftly procured him, and of courfe his emoluments and profits thereby, yet he hath ftill perfevered; wherefore, whatever approbation this production (of which there never has been any thing of the fort before in this kingdom, to which it is peculiar) may meet with, to him, in a very great degree, the honour is due. I did think it would have made two folio volumes in print, but abundance has been omitted which he, from his very far fuperior knowledge and judgment, conceived had not fufficient authority to fupport it.

At the fame time I muft obferve to my readers, that had I met with the fame indulgence from fome few other perfons in office, to whom, at
the

the time I have mentioned, I applied for inftruction and materials, as I did from thofe by whom
I had been favoured as aforefaid, the work might
have been advantaged; but I not only met with
refufals, but a furprizing ignorance, not only of
the bufineffes of their offices, but even of records
being in their poffeffion, which I knew to be
there, and which from my enquiries they afterwards found to be fo. One would not let me
take extracts without the order of government;
another conceived it would be difclofing the
arcana or fecrets of office, and this even in matters which it would be a public advantage, as
well that every man fhould be acquainted with,
as to inform the officers themfelves, and their
fucceffors. It is otherwife in England : and that
the matters or bufinefs of any public office (fave
thofe immediately connected with the ftate,
where fecrecy may be abfolutely neceffary) fhould
be concealed from the publick, feems nearly a
paradox.

Wherefore, I cannot help lamenting here, as
I have often done before, that there are not more
public officers in this kingdom who are lefs at-
tentive to the emoluments than to the knowledge
of the bufinefs of their offices; and that fkill,

abilities,

abilities, and true merit, hath at times been so little confidered in promotions to offices; and that the office had been ever fought out for the man, and not the man for the office. But of this I am fully convinced, that had the contrary been the practice in the revenue, and that the feveral officers employed therein, efpecially in the excife, had been raifed from one department to another, for their approved good conduct only, there would have been an exertion in all, not only in the doing of their duty, but to excel ; and the encreafed produce of the feveral revenues of the kingdom would at this day have been fuch, that not one half of the additional duties which are now in being, might at this day have been wanted. But where the falaries of offices are but fmall, (the moft of them being the fame as they were when one pound fterling of money would purchafe as much of the provifions of life as three will now,) and that ignorance, indolence, or de-merit, fhould be preferred to and have the fuper-intendency of knowledge, (which comes not by infpiration) integrity, and activity, than which, in the common tranfactions of life, there is not any thing more mortifying, if the poor difpirited, injured officer fhould in fuch cafe continue to be honeft, and merely execute his office, is it not

as

as much as may in reafon be expected of him?
I have often heard Sir Richard Cox, that very
able revenue officer, who died a commiffioner,
fay, that more than one third of the revenue of
excife of this kingdom was not then collected,
which might have been collected, and that much
was owing to this grievance, which for many
years before the time he mentioned this, had
been the common practice in the revenue.

If ever there was a likelihood of a thorough
reformation in this way, I think I may with confidence fay, it is in the prefent adminiftration of
this kingdom; where the moft intelligent, worthy,
generous, and accomplifhed nobleman, who pre-
fides in it, hath hitherto been (it is well known)
as induftrious to make himfelf acquainted with
the conftitution and bufinefs of the kingdom, as
he is to feek for, and reward the deferving; and
if not impeded by fome of thofe illufive fpirits,
(who, in this, as it is in another kingdom, under
the alluring fhew of patriotifm, or from a *little*
itch for popularity, a vice the inftant it is fought,
and is below the noble mind, are its greateft
enemies) would contribute cheerfully to make it
fpeedily a very flourifhing one.

The

The produce of the *cafual revenue* alfo, might beyond all doubt be improved feveral thoufands a year, and the execution of the public juftice of the kingdom (the far more material confideration) at the fame time forwarded greatly. This revenue arifes chiefly from forfeited recognizances acknowledged for the appearance and profecution of perfons guilty of breaches of the peace in outrages and violences of every kind; as alfo of fines and amerciaments, impofed by the feveral courts of juftice in the kingdom, on their officers and others for neglect or breach of duty in the execution of juftice, and for other offences.

Which recognizances, fines, &c. being eftreated twice a year into the Exchequer, are iffued thence twice alfo in the year, in the three feveral proceffes, commonly called the procefs of *green wax*; the firft, according to *magna charta*, againft goods only; the fecond againft body, goods, and lands; and third againft them alfo, and againft heirs, executors, and adminiftrators, (a full account of which is in this work) and directed and fent to the feveral fheriffs of the kingdom, in order to levy the feveral fums therein, for which they are to account in the Exchequer, after they are out of office, at certain times prefcribed by law for the purpofe.

Vol. I. B Now,

Now, were the attention paid to this very important department of the police of the kingdom, which, to promote its execution, ought to be paid to it, by thofe who are concerned in the feveral ftages thereof, befides increafing the cafual revenue very confiderably, it would be a principal means of promoting that due, that abfolutely neceffary obedience to the laws, which is fo much wanting in this kingdom, and of courfe contribute greatly to the prevention of the many riots and violences of every kind, for which it is at prefent noted above all the other nations in Europe; not one in twenty of which would happen, were it not for the ignorance of fome, the negleɑ or mifconduɑ of others, and, I much fear, the corruption of feveral among thofe who are employed in the conduɑ of this bufinefs, as I think I can prove to an abfolute demonftration.

For the purpofe, the evil originates often, indeed too often, with the juftice of the peace; who, being informed, either upon the *Examination* of the perfon injured, or the *Information* of fome other, of fome outrage committed againft the peace of fociety, perhaps negleɑs to take down the place of abode, occupation, or other addition, of the *Informant* or *Examinant*, by which
they

they may be afterwards found, either in the examination, information, or recognizances, to appear and prosecute; and if the offender happen to be brought before him, and that sureties are taken for his appearance, not only the same neglect or omission is committed, but persons taken as such sureties, who are neither of credit, substance, or known residence, how flagrant or outrageous soever the offence may have been.

Then, let the justices of the peace be ever so careful in these matters, and of transacting them properly, and that such additions of occupation, place of abode, &c. to ascertain the persons, shall have been inserted in their recognizances, yet, when these recognizances, either at the assizes, or at the quarter sessions, are on default of appearance or prosecution, ordered to be estreated, the several clerks of the Crown and peace of the kingdom (who are very material officers in this (I must again repeat it) most important business, and have it much in their power to promote or defeat it, neglect or omit to insert those additions in the estreats which are to be returned into the Exchequer, notwithstanding the rule of the court of the 22d of June, 1772, for the purpose. So likewise the same neglects or omissions are com-

B 2   mitted

mitted by thefe officers, where fines are impofed on tranfgreffors and defaulters, not only at the affizes and feffions, but alfo in fuperior courts of record, commiffions of oyer and terminer, &c. Nor is this all; many of thefe fines and forfeited recognizances, from favour, affection, party, partiality, or other improper confideration, are either not entered in the books of thefe officers, or, if entered, not extracted therefrom, or inferted in their eftreats; and often thefe eftreats are never returned.

But when thefe eftreats have been returned to the court of Exchequer, they are iffued in procefs to the feveral fheriffs of the kingdom, (as is before mentioned) to whom they are to be delivered by the purfuivant of the court, after he has received them from the feveral other officers thereof, whofe bufinefs it is to deliver them to him, and for all which tranfactions certain ftated times are appointed by rules of the court, that the fheriff may have fufficient opportunity, before the returns in the procefs expire, for the execution thereof, which has often happened otherwife, through the neglect or default, or other mifbehaviour, of the purfuivant.

Now,

Now, by the ſtat. 12 Geo. I. c. 4. ſheriffs ſhall have an allowance upon theſe accounts of 12d. out of every 20s. for every ſum not exceeding 100l. and 6d. for every 20s. over and above the firſt 100l. of all money (except poſt ſines) which they ſhall levy on the aforeſaid proceſs of the pipe or green wax proceſs; but this allowance is ſo greatly inadequate to the trouble and expence which muſt of courſe attend the collection of ſuch a number of ſums from ſuch a number of perſons, many of them wretchedly poor, and diſperſed through the whole county, that the high ſheriff leaves the whole tranſacting of this buſineſs to his ſub-ſheriff, who generally delivers the proceſs to the ſheriff's bailiffs to be executed, who are uſually of the loweſt of the people, and are not ſworn to the execution thereof; ſo that, perhaps, out of one thouſand perſons which may be in one of theſe proceſſes, it frequently happens that a ſheriff, on his appoſal in the Exchequer, may not account for ten of the ſums therein, (except cuſtodiam rents and poſt fines, in which caſes only the court will not receive ſuch a return, as the lands charged therewith cannot but be known, and the ſums of courſe be levied therefrom) but on the contrary,

poſitively

pofitively fwear, on the return of the bailiffs to them, that perhaps 950 perfons of the one thoufand have not either bodies, goods, or lands, in his county, although fuch perfons muft, in the cafe of recognizances, have appeared before the magiftrates who took the fame ; as alfo (frequently) in the cafe of fines in the courts, where they were impofed ; by which whole feries of mifconduct, this moft important and very expenfive procefs of juftice is rendered almoft totally fruitlefs, and his Majefty's cafual revenue is confiderably injured, to the real lofs of the public.

Quid trifles querimoniæ
Se non fupplicio culpa reciditur ?
Quid leges fine moribus,
Vana proficiunt ?

HOR. lib. 3. Od. 24.

But wherefore do we thus complain,
If juftice wears her awful fword in vain?
And what are laws unlefs obey'd
By the fame moral virtues they were made?

FRANCIS.

But now the queftion may reafonably be, what are the remedies for all thefe inconveniencies and mifchiefs?

In

In the firſt place then, it is propoſed, that by
a law to be made for the purpoſe, no perſon
ſhall hereafter be appointed a juſtice of the peace
in any county of the kingdom, who hath not an
eſtate of inheritance, or other freehold, or profit
rent on leaſehold intereſts, in the ſame county,
of at leaſt 300l. a year, except any number not
ſo qualified, not exceeding four at a time, for
the county of Dublin, to be approved of and
appointed according to the preſent method for
that purpoſe, and that they only be appointed
who are moſt noted for their abilities, wiſdom,
and integrity. As to their qualifications, ſee
1 Ed. III. ſtat. 2. c. 16. Eng. 34 Ed. III. c. 1.
Eng. 13 Ric. II. ſtat. 1. c. 7. Eng. 2 Hen. V.
ſtat. 2. c. 1. Eng. and 18 Hen. VI. c. 11. Eng.
none of which have been repealed, and are of
force in Ireland.

And as by ſeveral ſtatutes, alſo, many offences
are appointed to be tried by the juſtices of the
peace of the ſeveral counties of the kingdom, at
their quarterly ſeſſions, as if the ſame were regularly and properly held, ſuppoſe a law ſhould be
made for the more regular and effectual holding
of theſe ſeſſions, and for puniſhing by fine, or
removal,

removal, fuch of the faid juftices as fhould abfent
themfelves therefrom, without fufficient excufe ;
as this would be a certain means of ridding the
juftices of affize and goal-delivery of a confiderable part of that trouble which they have in
the trials of inferior, petty offences, the undoubted duty of the juftices of the peace, but
which at prefent is almoft totally and fhamefully
neglected by them, and enable the judges of
affize the more effectually to tranfact the more
material bufinefs of the country, as it would at
the fame time prevent that much to be lamented
lofs of time of the labourers of the country, who
are too much difpofed to be wantonly idle, in
attending the affizes feveral days, at the material
feafons in the year of ploughing, fowing, and
reaping, befides the coft and expences they are
at, moft heavy to them. As to the powers of
the peace juftices, fee the before-mentioned ftat.
34 Ed. III. c. 1. Eng. 36 Ed. III. c. 12. Eng.
2 Hen. V. ftat. 1. c. 4. Eng. and 4 Hen. 7. c. 12.
Eng. none alfo of which have been repealed,
and are likewife all of force in Ireland.

To reduce, then, all thefe ftatutes into one
fufficient and effectual act, for the regulation of
this moft important office, and to impower the
juftices

juftices of the peace alfo, at their quarterly feffions, to try and finally determine all demands or actions whatever, not exceeding 40s. value; and for that purpofe, and for the accommodation and eafe of themfelves, and the other inhabitants of the county, to hold each quarterly feffion at a different town in the county, and four of the principal ones to be fixed for that purpofe; and to have the fame fees upon all fuch actions as are now paid in fuits in civil bills; I am convinced I may venture to pronounce with certainty, that after a very few years perfeverance in thefe matters, as alfo in a cautious taking and due and faithful returning of recognizances, (than which there is not any part of the bufinefs more material) together with the neceffary affiftance of a properly appointed and well regulated office of fheriff and its under officers, fuch an appellation as a White boy, an Oak boy, or an Heart of Steel, would not be heard of in a county in the kingdom; and that good order, peace, induftry, and profperity, would be eftablifhed on a fure and permanent foundation in all. In England, although this office is not attended to or executed as it was formerly, and as it ought to be, yet it is far better there than it is here. That juftices of the peace may (as well as fheriffs, clerks of the

Crown and peace, and others) be punifhed for their neglects or mifconduct in this bufinefs, by fines impofed on them by the court of Exche-- quer, there have been inftances; or the chancellor may fuperfede the commiffion and remove the perfons fo charged from the office, upon the mat- ter being properly laid before him by the Barons of the Exchequer; but it is far better to prevent a mifchief than to have occafion to punifh the offender.

The officer next in order, to be confidered, and a very principal one in the conduct of this bufinefs, is the fheriff of the county: he is (as has been juft mentioned) the collector of this branch, as he formerly was of all others of his Majefty's revenue, and as fuch is entrufted with the execution of the green wax procefs; and on his fidelity and diligence therein depends in a great meafure the advantage which is to arife therefrom to the publick, in the prefervation and fecurity of the peace and happinefs of fociety, and the improvement of this branch of the King's revenue, which for the benefit of the publick, is applied in aid of other revenues to defray its expences.

Wherefore,

Wherefore, it is to be wifhed that none but the principal gentlemen of the firft reputation and credit, with fufficient eftate, (at the leaft 500l. a year) in the feveral counties in the kingdom, were to be appointed to this office of dignity, truft, and authority; upon the due and proper execution of which, the property, the life, and the liberty of every individual, and the peace and fafety of the whole, abundantly depend, and for the defence of all which it was beyond all doubt originally inftituted, as may appear by the many excellent laws which have heretofore from time to time been made for the appointment, qualifications, and powers of this great officer; fuch as Artic. fuper chart. 28 Ed. I. c. 8 & 13. Eng. 9 Ed. II. ft. 2. Eng. 2 Ed. III. c. 4. Eng. 4 Ed. III. c. 9. Eng. 5 Ed. III. c. 4. Eng. 14 Ed. III. c. 17. Eng. and 12 Ric. II. c. 2. But the high importance of this office will beft appear from a relation of fome of the effential bufineffes with the execution of which this officer is entrufted, for the advantage of the community.

Does he not, then, return the juries who are to try our lives, our properties, our liberties? And if in this he is corrupt, would not this main

C 2   pillar

pillar of our freedom, this moſt valuable bleſſing, (which of all the people on the globe they of our glorious conſtitution only enjoy) be as tho' we had it not? Is it not by this officer that our laws are to be finally executed? And if in this he is corrupt, (which from men of ſcanty circumſtances, or ſmall reputation, there is but too much reaſon to fear might be the caſe) would our laws be then any other than a mere dead letter, to the utter de-ſtruction of credit and commerce? But above all, is the high truſt which is repoſed in him on elec-tions of members for the legiſlature; on his con-duct in which, our glorious conſtitution chiefly depends *.

At the ſame time I muſt obſerve, that if any proceedings have happened in the appointment

* To ſuch a pitch is the abuſe of this very eſſential office at preſent arrived, that it is twenty to one if a writ be executed in any of the diſtant counties of the kingdom ; or if it be that the plaintiff is not by the iniquity of an under-ſheriff kept from the benefit of it until he is more wearied in ſeeking it from this ſub-ordinate miniſter of juſtice than he was from his original debtor, to the almoſt entire deſtruction of credit; which verifies what has been ſaid of our conſtitution, " that we of all civilized nations have " the beſt framed, but worſt executed laws." Nor is the Crown's revenue, ſuch part as the ſheriffs collect, as aforeſaid, more eaſily got out of their hands.

to this office, in the leaft degree inconfiftent with the aforefaid feveral ftatutes, or with the ftricteft adherence to the eftablifhed principles of our conftitution therein, it muft in this cafe have arifen (as it is well known it did on the Excife law,) from an abfolute neceflity, and from this unaccountable miftake, that the intereft of the Crown and that of the people can, in the true and real fenfe of the matter, in our conftitution, poffibly be inconfiftent; from which it has often been as difficult to get fheriffs to return jurymen where the Crown has been concerned againft the fubject, who were not biaffed in favour of the latter, as it was alfo to get jurymen who were not fo biaffed where fheriffs were really impartial; wherefore, the complaints of grievances, which, it is alleged, have arifen to the publick upon thefe occafions, are in the general bellowed out by thofe, who, too often, from their own improper proceedings, have been the caufe of any alteration, or change, which may have been in the proceedings in either of the two departments I have mentioned *.

But

---

* The method which has been for many years of appointing fheriffs is thus : The judges of affize, on their fummer circuits, require the fheriffs in office in the feveral counties in the kingdom, each of them, to return them the names of three perfons in each county

But to return, if fheriffs upon the execution
of this procefs, inftead of 12d, which is all they
are now allowed on their accounts, as aforefaid,
out of every 20s. for every fum not exceeding
100l. were to be allowed 5s. and inftead of 6d.
for every 20s. over and above the firft 100l. to
be allowed 2s. 6d. for all money, (except poft
fines and cuftodiam rents, in which laft cafes to
be allowed 6d. in the pound only) it has been
conceived, it would fpeedily have that moft falutary and much to be wifhed-for confequence, of
greatly fecuring and preferving the fafety and
peace of fociety, as the perfons fined or mulcted
in this procefs could not then afford to tamper
with the under officers employed by the fheriffs
in the execution thereof; which procefs in many
parts of the kingdom are, for the reafons I have
herein mentioned, abfolutely held in contempt;

county proper to fucceed them, which they accordingly do; and
at the meeting of the judges in the chancellor's chamber on the
morrow of All Souls in the following *Michaelmas term*, the lord chancellor calls on them for their returns, which, when received, he
delivers to the lord lieutenant, who *appoints* one for each county
out of each return. But note, the judges have a power before they
make their returns to alter the perfons, or any of them, in their
difcretion. All which is a good deal agreeable to the aforefaid
ftat. 12 Ric. II. c. 2. But fee Blackft. vol. I. 339, &c.

fo

fo that our laws are quite ufclefs, mere *bruta fulmina et vana*, befides the lofs of feveral thoufands a year to the cafual revenue; for even after the allowance of this large poundage to fheriffs, the encreafe to this revenue would be confiderable from the collection of innumerable fums which in all likelihood, for the reafons before mentioned, might never have happened. What induced the lords to reject the bill for this purpofe, the laft feffion, after it had been approved of in all the other ftages through which it had paffed, is hard to conceive, unlefs it was occafioned by a few miftakes that were in it, which had been introduced in fome alterations which had been made in the original draft, which was prepared by me.

At the fame time, the judges, who have the difcretionary power of impofing fines for offences *unaffeered per pares*, are ever to bear it in mind, that in an Britifh conftitution, an abfolute neceffity only can warrant it; and that of whatever benefit the *ufe* of it may be to the publick, in the ways I have mentioned, yet, that its *abufe* might caufe us to wifh it had never exifted. To confider alfo, that no plea can be to the Eftreat of a fine which is not firft laid before the *court of Exchequer* for their permiffion. The recogni-

zances

zances ſtand upon a different footing, they are acknowledged by the parties thereto.

Beſides, as to this office of ſheriff, this moſt important as laborious office, for the ſervice of the publick, it is a matter well known, that what from the expence which ſheriffs are at in the paſſing of their patents, the heavy charges on them in ſeveral of the offices of the Exchequer on paſſing their accounts, but above all on re-ducing of fines, which have been impoſed on them by the court, for the neglects, defaults, or other miſbehaviour, or miſconduct, of the under officers, which they muſt of neceſſity employ, and chiefly in this buſineſs of the green wax proceſs, they are, in the general, conſiderable loſers by the office; and that not only they them-ſelves, but their families after them, have often been involved in the moſt diſtreſſing difficulties.

Even the indulgence they receive in the length of time which is allowed them by the court *to account, to pay their tots*, and *to clear their ac-counts*, (the three ſtages through which they are to paſs in order to be diſcharged and obtain a quietus,) being at leaſt double what it was formerly, as will appear by the books in the

treaſurer's

PREFACE.

treafurer's remembrancer's office, (which alteration, and the miftake on which it is founded, is fully fet forth in the following work) whereby, before the fheriffs are compelled to account, their fub-fheriffs and their fureties may be rotten in their graves.

It has been the opinion of feveral of the firft in knowledge in this branch of the revenue, that the prefent courfe of the procefs of green wax might be abundantly abridged; or that one well-ordered writ might do inftead of the three which now iffue, whereby a prodigious expence would be faved to the Crown, as alfo very great labour and lofs to the fheriffs of the kingdom, and much benefit gained to the publick: whereas, by the prefent courfe, the *Pipe*, or fecond procefs, without any colour of reafon, and againft a ftanding rule of the court (23 Nov. 1685) to the contrary, iffues twice; and then the *Treafurer's remembrancer's procefs*, with the *Prerogative writ*, alfo called the *long writ*, annexed thereto, which, although it be againft every thing, body, goods, lands, heirs, executors, and adminiftrators, and an *Inquiry to be held thereon*, yet is rather lefs effective than any of the others; nor is it to be wondered at, from the manner in

VOL. I.   D   which

which it is executed, which is thus; the fubfheriff in fome little town or village in the county,
perhaps in a cabin on the road, where twelve of
the loweſt of the people, his bailiffs and others
his creatures, are the jury, and a general inquifition returned of neither body, goods, lands,
heirs, executors, or adminiſtrators in the whole
county, as to every perfon contained in the procefs, although by a ſtanding rule of the court
(14th May, 1717,) the inquiry on this writ is to
be held in every barony in the county; fo that
in a courfe of thirty years, for which period I
have been Solicitor for the cafual revenue, I have
not feen as many fums brought in thereby; and
the fame nugatory proceedings have been of late
upon the Pipe procefs. But as all thefe matters
may be the better judged of from the whole of
the proceedings, which are in the following work,
with my occafional obfervations thereon, I ſhall
refer my readers thereto.

Others have thought, that it might anfwer the
purpofed ends much better, if the prefent mode
of collecting the cafual revenue was to be changed,
as that of the other revenues hath been, by transferring it to the feveral collectors of Excife in the
feveral diſtricts in the kingdom inſtead of the
fheriffs,

fheriffs, with good allowance on collection, but
not by falary; and efpecially as the books of the
hearth-money collectors would be of fingular ufe
therein, by which the places of refidence of the
inhabitants of every county, who pay hearth-
money, might eafily be known : befides, copies
of the fummonifter's procefs are fent out twice in
every year, to wit, in Hilary and Trinity vacations, by the Solicitor for the cafual revenue *
to the feveral collectors of the kingdom, to en-
quire of the feveral perfons therein, and of their
perfonal effects, and to make return thereof to
him, or to the commiffioners of the revenue,
that he may therefrom be enabled to cheque the
feveral fheriffs on their accounts upon this procefs; fo that they have already a confiderable
part of the trouble which they would have, were

* He is by this office fuperintendent of every other perfon
concerned in the bufinefs of this branch of the revenue;
and fhould therefore with the greateft attention and diligence
purfue the *Inftructions* he receives with his commiffion. This
Office, it is to be obferved, was formerly held with that of the
Clerk of the Informations in Dublin port, under the fame comiffion,
which was worded as if the bufineffes of both were connected;
whereas, no two in the revenue are more foreign to each other; and
they were alfo held by perfons ignorant of the law, whereas attornies only are the perfons proper to conduct them, as the aforefaid
Inftructions for both, (which are blended, but may be eafily feparated,
and annexed fo to the commiffion for each,) will fully evince.

D 2   they

they to be the collectors of this branch of the revenue. It is a matter very worthy of confideration, and would require the matureft. That it would deprive feveral officers of the Exchequer of large fees and profits, which they make by this branch of the revenue, and by the accounts of sheriffs, is moft certain; but this is a matter, which, if put in competition with the advantage of the publick, is fcarcely worth a thought, as they may be recompenfed by the publick, and efpecially as by the prefent mode of proceeding, a large expence is incurred with but little profit to the publick, whereas the advantages to it would be exceeding great were this bufinefs properly conducted.

The clerks of the Crown and peace, as well as the juftices of the peace of the kingdom, have it equally in their power to promote, as to poftpone, or defeat the execution of its publick juftice, as alfo to improve, or reduce the income of the cafual revenue. For inftance, if juftices of the peace would be careful in not fuffering any perfon to become fureties for the appearance of perfons charged with offences, but fuch as are of fome degree of credit and fubftance, and known refidence; if they would infert in the recognizances

which

which they take, the places of abode and occu-
pations of fuch, as it will appear by the following work they are bound to do, or may be fined,
nay, removed; if the clerks of the Crown and
peace alfo would be as careful to do the like in
their Eftreats, which they return to the Exchequer, and as punctual in the returning thereof as
they are alfo bound, under the like penalties, to
do; if on all fines hereafter to be impofed in
any of his Majefty's courts of record in Dublin,
or elfewhere, commiffions of oyer and terminer,
as alfo at the affizes or feffions, and other courts,
where fines or amerciaments are ufually laid or
impofed, the feveral clerks of the Crown and
peace, or other proper officer, would immediately enter down the additions and places of
abode of the perfon or perfons fo fined, (all
which requifites the faid feveral officers alfo, by
the rules of the court, as will appear by this
work, are bound to perform) and that the judges
of affize would themfelves compare the Eftreats
with their own private court-books, which they
fhould ever keep as a check upon thofe of the
clerks of the Crown, and in which they fhould
be moft careful to enter every forfeiture and every
fine they impofe, in order to prevent the groffeft
offenders (from any intereft or improper influence,

as

as has been often the case) from escaping the punishment they justly deserve; and to see that true and faithful returns are made thereof, or else this most important and expensive procefs must become a perfect nullity.

Then, the offices of clerk of the Crown and peace are usually in the same grant, through the whole kingdom, so that the deputation of the latter is as usually purchased or farmed; the probable consequential evils of which are so glaring, it were needless to suggest them : wherefore, it has been conceived, it would be better for the publick if the offices were to be separated.

But there is another matter which also is most worthy of attention. In England, by several statutes there, 37 Hen. VIII. c. 1. 3 & 4 Ed. VI. c. 1. and 1 Will. & Ma. stat. 1. c. 21, the *custodes rotulorum* there have a power of appointing clerks of the peace, yet notwithstanding that these offices, in this kingdom, are by the King's grants, and that there is no such statute here, yet several lords lieutenant of counties (as they are here cal-led) have taken on them to appoint to this office of clerk of the peace, which is not only most injurious to the legal patentee, (who is generally the

the purchafer thereof) but alfo of much mifchief
to the publick.

Then again, there are feveral corporate cities
and towns in this kingdom to whofe corporations,
all the fines, ranfoms, and amerciaments, for all
crimes and trefpaffes within fuch cities and their
precincts, and all recognizances, penalties, and
forfeitures, of all the citizens and inhabitants
therein, are granted; who therefore have conceived that the clerks of the Crown and peace, of
their jurifdictions, are exempted from returning
the eftreats thereof to the court of Exchequer,
and orders of the court have been inconfiderately
(I believe) conceived to that purpofe. But the
better opinions feem to be, that thefe fines, &c.
ought to be eftreated for the fake of publick juftice, as alfo of the party on whom they were impofed, who if he conceives they were illegally or
improperly impofed, may, on application to the
court, be admitted to plead to the eftreat thereof,
or they may be reduced, if foundation for favour
fhould appear to the court, and partial proceedings prevented, where thefe indulgent grants are
given.

And

And furely alfo, never did a fairer opportunity
offer than in the prefent adminiftration, to petition for fome law to reftrain the daily practifed
abufes and confequential grievous mifchiefs which
attend the *obtaining cuftodiams upon outlawries in
civil actions between party and party*; which are
injuring every day more and more the common
fecurities of the kingdom, and deftroying its
credit; and all this moft abfurdly under the fiction of the prerogative of the Crown, which is
no more really concerned therein (as I have before
mentioned in the preface to my treatife on the
Pleas-fide of the Exchequer, to which my readers
are referred) than is the prerogative of a *Nabob of
India*; yet, were it fo, it was even faid by King
Charles I. (who fell a facrifice to his zeal for
what he thought the prerogatives of the Crown,
and the rage of fanaticks) in his anfwer to the
petition of rights, in the third year of his reign;
" That his prerogative was to defend the rights
" and liberties of the fubject, as were the rights
" of the fubject to ftrengthen his prerogative."
Will any perfon, then, be hardy enough to fay,
that this prerogative fhould ever be in fiction
ufed to injure thofe rights? And would not
fuch a proceeding be rather an injury than an
advantage to the prerogative? And is it not an
offence

offence to our virtuous, pious Sovereign, whofe
benign heart, I am convinced, it would grieve,
were he to be apprized of the unfit ufe which
is fo frequently made of his royal prerogative. But as I have in my faid former preface
endeavoured to fet forth the whole of the very
improper, nay, unjuft proceedings on this procefs, I fhall only fum up here the many grievances
which are the fure attendants thereon; nor fhould
I have repeated any of them, but that they more
properly belong to the following work, as alfo
the more ftrongly to inculcate them on the minds
of thofe who may procure the redrefs.

This procefs, then, which is againft body,
goods, and lands, (and by which the unfortunate perfon againft whom it iffues is proclaimed
a contemnor of the laws, a rebel, and a fugitive,
although not ferved with any procefs, fummons,
or previous notice thereof whatever, and is vifible
every day) may be iffued for the fmalleft fums,
for an uncertain, nay, for a fictitious fum, (as is
often the cafe) for neither bond, judgment, note,
affidavit, or other voucher is produced, or even
required, as it is in every other cafe, as a foundation for this violent proceeding; and yet it
fhall have preference to, and take place of every

other procefs in the law, even in cafes where the moft folid fecurity has been given.

The dowered and the jointured widow, the purchafer, the judgment creditor, the mortgagee, unlefs he be in the actual receipt of the rents *, and the every other real, fair incumbrancer, are often without the leaft notice (for almoft the whole proceedings are, as I have fet forth in my faid preface, as clandeftine as they are injurious) ftript of their fecurities, put to great, to grievous trouble, and to moft unjuftifiable · expence, the coft having been often many times more than even the fictitious fum ; the *before* miferable tenants of the eftate eternally harraffed, until driven to emigrate. The landlord, fhould it be a derivative intereft which is attached on this procefs, rendered incapable of bringing an Ejectment for non-payment of rent without the permiffion of the Exchequer, on a motion to be made by his counfel; (which permiffion even cannot be applied for, without the confent of the Attorney general, as guardian of his Majefty's prerogatives, to be

---

* But quære, If the mortgagee be not in the actual poffeffion, on default of payment of the money on the day appointed by the deed, and if the mortgagor be not thereby abfolutely tenant to the mortgagee for the lands.

previoufly

previoufly had for the purpofe) whereas, this unfortunate landlord may have no more to fay to the debt or demand than an inhabitant of Siberia, and muft be a diviner to know the names of the many creditors of his tenant who have cuftodiams againft him, as otherwife the fearch for them may be endlefs, befides the heavy expence attending this motion, (for which fee my faid preface, and the chapter on Cuftodiams in the following work) and all this under the aforefaid fiction of the Crown's prerogative being concerned therein.

Then, this fame moft unlucky outlawed perfon (who perhaps on a fair trial might not owe the plaintiff a fhilling) is thereby put out of the protection of the law, fo that he is incapable of fuing for his rights, or bringing any action for redrefs of injuries, and all his goods and chattels, without any committed offence, forfeited to the King; he is incapable of being either a grand or a petty juryman; and fome have thought (which I leave to the learned) of voting on an election, or filling any office in the ftate, whilft this (perhaps moft unjuft) outlawry is fubfifting againft him.

E 2   And

And then, upon this proceeding, the creditor takes all, and by *Elegit* but a moiety; by this, he gets an *actual poſſeſſion*, on the ſame fiction that the King is concerned; by *Elegit*, in the general, but a legal poſſeſſion. Suppoſe then, that this procefs ſhould never be permitted to iſſue but upon a poſitive affidavit of at leaſt twenty pounds being juſtly and fairly due by the defendant to the plaintiff; that no priority ſhould be given to the execution thereon, but as it is now by law between *Elegit* and *Elegit*; and (as I have before mentioned) that *Elegits* on judgments ſhould reach the whole eſtate, as is the caſe on theſe Cuſtodiams, and on ſtatutes of the ſtaple; and that the fair creditor ſhould get an actual poſſeſſion thereon, without the trouble, loſs of time, and expence of an Ejectment to get a ſecond poſſeſſion of the ſame thing; or provide (as in other caſes) that all the aforeſaid ſtrongly intereſted perſons ſhould have real and ſufficient notice of the oppreſſive and diſtreſſing proceedings on this procefs; would it not be better for all the perſons I have mentioned as intereſted, which may include the whole nation, except the Attornies and Officers of the courts who iſſue, and the ſheriffs and ſub-ſheriffs who execute them. It has been

laid,

said, that the judgment creditor may prevent his being injured by this procefs, by extending immediately on an *Elegit*; but every man who knows any thing of thefe matters, muft know that this would as effectually deftroy this common fecurity of the kingdom, as this procefs is effecting it every day.

Attempts have been made, and fome of them of my promotion, to relieve us from the dreadful *abufes* which attend the proceedings on this procefs: but, alas! the private intereft of a few individuals prevented it, as hath been too frequently the cafe in this unfortunate kingdom; yet, old as I am, I will not yet defpair of feeing it accomplifhed; if not, I have this comfort, that I have done my part to the beft of my abilities, and without any private gain or felfifh view. The paucity of cafes in the books (I mean the Englifh publications, for there are none here) will evince how very fparingly the procefs of outlawry have been ever ufed in England; and in one of them, 12 Mod 413, there is a cafe, where a perfon having outlawed another in a civil action, whom he knew was vifible, and might be eafily ferved with procefs, was ordered to reverfe the outlawry at his own charges. And fome years ago, having written to an officer of the Exchequer in England,

England, my acquaintance, for the proceedings upon *Custodiams*, and *Injunctions* upon such Out-lawries, in civil actions between party and party, he sent me the rules of the revenue side of the court of Exchequer there, (which are yet in my poffeffion) and among them, there is but one in any fort relating thereto, which is of the 13th of May, 16°9, and is, " That where any out-" lawry shall be tranfcribed into the court, and " procefs made out thereon, and afterwards such " outlawry shall be reverfed, before any judgment " shall be entered for removing the hands of the " Crown, and the party outlawed reftored to his " poffeffion, the profecutor of such outlawry shall " be paid such cofts as shall be taxed by their " Majefties Remembrancer, or his deputy, for the " proceedings in the said court." And at the fame time informed me, that he had been feveral years an Attorney of the court, and had never been concerned in any such proceeding. And in 2 Atkins, 408, it is mentioned, as if it were a fpecies of proceeding peculiar to this kingdom. In truth, it is a proceeding the *abufe* of which is almoft equal to the deftroying of the credit of a country, and is a difgrace to the juftice thereof.

There

There is yet another object worthy of attention, and that is the present state of the malt liquors of this kingdom, that indifpenfible neceffary of life; a matter worthy of moft ferious confideration and attention, not only with regard to the great lofs which the kingdom fuftains by the prodigious fum which the vaft importation of this commodity takes from it yearly, and the great diminution of his Majefty's revenue of Inland Excife; but chiefly to the health and morals of the lower fort of people, which are almoft deftroyed by the fubftitution and too general confumption of fpirituous liquors; there being hardly a village, nay, even a large town or city, in the kingdom, where a drop of Irifh Ale or ftrong beer is to be had which the pooreft wretch can with fafety admit into his ftomach; nothing but Englifh Porter, which they of circumftances only can purchafe ; and even this from the flownefs of its vent, and the adulteration it fuffers, is often hardly drinkable, to the extreme great prejudice of thofe two important objects, tillage and manufactures.

In order the more fully to inveftigate this moft interefting matter, it will be neceffary to take an account, which may be eafily obtained, of the

quantity

quantity of Englifh ftrong Beer and Ale imported at prefent into this kingdom; of the duty of cuftom and excife, and other charges here thereon; the Inland Excife upon the Irifh malt liquors; the expences of this manufacture; the difference in thofe expences between the time the duties were impofed thereon and now, and the prices at which the Englifh and Irifh malt liquors are in the general fold at in both kingdoms; in which, we are alfo to take into confideration, the great advantages which the Englifh brewers have over thofe of this kingdom, from the prodigious bounty upon exportation; in the meafure of their gallon, and in the hops to the brewers there and here, not only in price, but in the quality, as they have the firft choice, &c.

When all this fhall be done, I am inclined to think that, on a fair comparative view, it will appear that until the brewers here have fome fur- ther encouragement for brewing better liquor than they have at prefent, we may defpair of ever having it; as fome proof whereof, it muft be within the memory of many, when good wholfome ale was fold in this city for two pence and excellent for two pence half-penny a quart; fo that it was ufual for mechanicks, and others of higher rank,

to

to fpend an evening over a cup of it; and there were feveral Brewers then extremely wealthy ; whereas, now it is much otherwife, though there is not near the number in the trade that were then.

But now, as to the benefit which may arife to the revenue of this kingdom from the encouragement and improvement of the Brewing trade of it : The ftrong Beer and Ale imported here from England laft year, amounted to near 54,000 barrels, of which thofe of ftrong Beer, called Porter, were not lefs than 53,000, for the importation has been encreafing every year for fome years paft. For thefe 54,000 barrels about £56,700, Irifh money, has been fent from this kingdom, at 21s. Englifh, a barrel, which is what the importing Irifh merchant only pays for it, though fold for perhaps 30s. for confumption in England, as he has the benefit of the drawback or bounty in England, upon exportation from thence ; fo that what with £9,000 being the freight thereof, at 20s. a tun, as alfo the coft of infurance, and other charges, as for butts, hogfheads, barrels, &c. not a lefs fum than £70,000 is fent from this unfortunate kingdom, and muft increafe if not prevented by brewing better liquor here.

The duties of Excife and Cuftom, then, upon
the quantity fo imported, according to the prefent method of rating it on the contents of the
Englifh barrel, which being fome fmall matter
under 1s. 8d. a barrel, amounts to very near,
but not quite, £4,500 per annum; (which imported Beer is not liable to any Inland Excife)
whereas, the duty upon the like quantity manufactured here, which, from the late alteration
made in the meafurement of the gallon, refpect-
ing the additional duties, and the lofs of Excife
thereby, is at the rate of 4s. and fome fmall
fraction of a halfpenny, per barrel, would
amount to upwards of £10,800; fo that in this
cafe, the increafe to the revenue of the kingdom
would be £7,200, which would increafe with
the confumption; befides the increafe in the revenue (and no inconfiderable fum) by the duty
on the additional quantity of hops ufed in fuch
brewing, which would be fupplied from England.
And then whatever detriment the prevention
hereby of the importation of fuch Englifh malt
liquor might be to fome individuals in England,
it would be no lofs, nay, it would be a faving
to the revenue thereof, as not only the whole
                                            Excife

Excife thereon, being 8s. a barrel, is drawn back upon the exportation of thefe liquors, but 1s. premium given upon every barrel fo exported, when barley is at 24s. a quarter, or under.

It has occurred to fome perfons (if it could be done without injuring his Majefty's hereditary revenue, which fhould never be infringed whilft we regard the prefervation of the conftitution of this kingdom; an affertion which may feem ftrange to fome, yet is moft certainly true,) in order to encourage the Brewers, to take off fome of the duties upon malt liquors, and to lay an equal portion upon malt. This, on the other hand, has been objected to, as it might be fubjecting the country gentlemen to the Excife laws; but there is fuch a duty in England without any fuch inconvenience, for by the Act which induces it there, a compofition may be made for it; and fo it might be here.

It is true, the Brewers of this kingdom have lately had a very great advantage in the alteration which has been made in the meafure of our gallon, to wit, from $217\frac{6}{10}$ to $272\frac{1}{4}$ cubical inches, fo far as it reaches, which is only to the

F 2 *Additional*

Additional Duties, not to the *old Excise*, it being the King's *Hereditary Revenue*, and not to be infringed without his previous confent; the hiftory of which alteration is as follows.

Sir James Shean and partners, to whom, and ten others, the Revenues of this kingdom had, in the year 1676, been farmed at the yearly rent of £240,000, having obferved that the gallon by which their predeceffors * had received the duty of excife, did contain 272 $\frac{1}{4}$ cubical inches, when, at the fame time, the common gallon ale meafure, made ufe of through the kingdom, and which was authorized and fealed by the feveral clerks of the market, did confift of no more than 217 $\frac{6}{10}$ cubical inches, being juft $\frac{4}{5}$ of the gallon by which they received the duty; and this being a lofs to the *Farmers*, upon enquiry how the law was as to this point, they found that by an Irifh act, 28 Hen. VI. c. 3. it is enacted, " That there fhall be but one meafure throughout the kingdom, that is to fay,

* To wit, John Forth, of London, alderman, to whom, and ten others, the faid revenues had before, in the year 1669, been farmed for feven years, at the yearly rent of £219,500, as appears by the deed in the *Rolls*.

the

the gallon, the pottle, the quart, the pint, and the half-pint, for meafuring wine, ale, and other liquors;" but it does not mention what the contents of the gallon ought to be.

That in another Irifh act, 7 Will. III. c. 24. there is a gallon for meafuring corn appointed, containing 272 ¼ cubical inches, which is anfwerable to the Winchefter meafure; this meafure was to remain in the Exchequer for a ftandard.

The 10 Hen. VII. c. 22. makes all the laws in force in England to be fo in Ireland, the *Farmer* then enquired how at that period the gallon was afcertained in England, which was as follows, by the 51 Hen. III. ftat. 2. the gallon was thus fettled, an Englifh penny, called a *fterling round,* and without clipping, to weigh 32 wheat corns in the midft of the ear, 20 pwts. to make one ounce, 12 ounces one pound, and 8 pounds one gallon of wine, and 8 gallons of wine to make one *London* bufhel, and 8 bufhels one quarter.

In the 12 Hen. VII. c. 5. Eng. all the meafures in England were called in, and a new ftandard meafure was erected; and as before the faid ftat.

the

the gallon was to contain 8 pounds of wine, the
gallon thus eſtabliſhed was to contain 8 pounds
of wheat, of Troy-weight; and as 8 pounds of
wine put into a cavity that ſhall juſt receive it,
and no more, is to another cavity that ſhall contain 8 pounds of wheat, ſo is 217 $\frac{6}{10}$ cubical
inches, the contents of 8 pounds of wine, to
272 $\frac{1}{4}$ cubical inches, the contents of 8 pounds
of wheat.

Thus the *Farmers* found out the contents of the
liquid gallon in England in the 10 Hen. VII.
and at all times before; and as the Iriſh uſed that
gallon in the common ale meaſure, and as the
Engliſh Act of 12 Hen. VII. was not made an
Act in Ireland, they conſidered the gallon containing 217 $\frac{6}{10}$ cubical inches, as the only legal ſtandard
meaſure in Ireland, and upon a controverſy heretofore between the *Farmers* of the Revenue, when
it was ſet to farm, and the Brewers, the matter
was debated at a Council board, who gave their
opinion in favour of the gallon 217 $\frac{6}{10}$ cubical
inches.

The practice of taking the duty of Exciſe
upon Beer and Ale, by the ſmall gallon of 217 $\frac{6}{10}$
cubical inches, continued until the 11 & 12 Geo. III.

c. 1.

c. 1. when the legiflature, upon confidering the act of 12 Eliz. c. 3. and the faid ftatute, 7 Wil. III. c. 24, which mention the ale gallon as containing 272 $\frac{1}{4}$ cubical inches, enacted in the money bill, that the Excife fhould be taken by the gallon containing 272 $\frac{1}{4}$ cubical inches. The difference, then, between 217 $\frac{6}{10}$ and 272 $\frac{1}{4}$ is a fifth part, fo that the Crown lofes almoft a fifth part, or near 1s. of the additional duty formerly received, but it is in this only, for the alteration is not ufed in taking the *Old Excife*, (although it be claimed by the Brewers) for the reafons I have mentioned before *.

I am

* When Humphry French was lord mayor, an Act paffed, 9 Geo. II. c. 19, §. 1. making the Dublin Brewers barrels to contain 40 gallons, and the half barrels 20 gallons, according to the 217 $\frac{6}{10}$ cubical inches, which accounts for the above alteration, that each gallon of the barrel of 32 gallons fhould contain 272 $\frac{1}{4}$ cubical inches ; for 32 gallons of the latter are exactly equal to 40 gallons and fix tenths of the former dimenfions.

The Dublin Society once had the moft important object of encouraging by premiums or bounties the improving the malt liquors of the kingdom, which, at the fame time, would alfo much promote its agriculture, (the primary caufe of that noble inftitution,) greatly at heart, as their then proceedings will fhew ; but a recent
rage.

I am convinced, if the Revenue laws of this kingdom were now to be collected, contracted, and properly digefted, and every law by which a forfeiture, or a penalty might be incurred, were to be promulgated as publickly throughout the kingdom, as poffible, it would be productive of much advantage to the Revenue, not only in the prevention of many frauds, but the faving of a confiderable portion of the expences attending the many legal profecutions on account thereof. The collections of fuch of thefe laws as had been made antecedent to the 33 Geo. II. c. 10. by which the number mentioned in the ftatute of the 31ft. of the fame reign, c. 6. with thefe ftatutes, and the feveral laws fince made, form at this day an abfolute mafs of contradiction, confufion, and perplexity. And there is not a feffion of par-liament in which there are not as many new Acts, or claufes for Acts, propofed as there are under officers, who wifh to have as little trouble in their

tage in a few (who feldom fail to attend, and influence others) for certain manufactures which we cannot export, (I need not fay more) hath fo much engroffed their attention of late, that *Agriculture* is become too much but a fubordinate confideration; and the reclaiming the *Waftes* of the kingdom, (many millions of acres) thofe inexhauftible mines of population, wealth, and ftrength, (of which an Earthquake, or fome fuch fpecial vifitation of Heaven, only could deprive us) almoft totally flighted.

employments

employments as poffible. It would be well worth while to pay generoufly fome gentleman of the bar to complete what I have here mentioned againft the next feffion of parliament.

And now to conclude; the moft honourable and refpectable lift of fubfcribers to thefe my attempts, cannot but make me a little vain; yet I tremble for their fuccefs, and, with great humility, crave leave to hope, that whatever they may fall fhort of any expectations from them, that the moft favourable indulgence of my readers will confider, that they are the firft of the kind in this kingdom, nor have I met with, or ever heard of any upon the fame plan even in England; as alfo that the well-known multiplicity of bufineffes in which I have been all along engaged (perhaps exceeded by none that ever was of the fphere I am in) will be kindly taken into the account.

As to the omiffions in the body of the work, which are fupplied at the end of it, the fearching for the Rules of the Court was very laborious, as the office books were not indexed or alphabetted (as it is called) for near fixty years after the time from which I took my fearch; and after the firft volume had been printed, upon a fecond fearch, (which not till then occurred to me was neceffary) I found that fome had been omitted in the Indexes, or mif-indexed,, and one new one after made. But in order to make amends the beft way I can, I have, after the *general* Index to the work, given *a particular* fhort one of all the Rules only, of the Revenue fide of the Exchequer, which will be very fatisfactory to the Court, and the Practitioners. Then, other matters occurred as worthy of infertion fince the firft volume was printed, and as the work is divided into chapters, they could not then be inferted in thofe to which they properly belonged.

It is alfo requefted, that on receipt of the two volumes, the errata of the prefs (fuch as are material, for which I muft for the fame reafons alfo plead for indulgence) may be corrected with a pen ; half an hour might do it effectually.

But here I cannot help obferving, that not-withftanding the contents and propofals had been not only a confiderable time advertifed in feveral of the publick papers, but pofted up in the halls

of

of the Four Courts, and in several of the publick Coffee-houses in the city, yet, not even a single Merchant, nay more, not a Revenue Officer, save the Commissioners, their two secretaries, and about four more, appears in the list of subscribers, although the work so especially appertaineth to them and their respective businesses.

CON-

CONTENTS.

VOL. I.

CHAP. I.

CHAP. II.

CHAP. III.

CHAP.

C H A P.

CONTENTS.

CHAP.

C H A P. XVI.

C H A P.

CONTENTS.

C H A P. XXVII.

CONTENTS.

VOL. II.

H 2 No. 3.

CONTENTS.

Page

CONTENTS.

N. B. Lord Chief Baron Parker's Reports of
Revenue Cases in the Exchequer in England
having appeared since the two Volumes of this
Work were finished, and it being apprehended
that a few points therefrom may be acceptable,
they are accordingly inserted after page 345 in
the second Volume.

⁎⁎ For the Errata see the End of the second Volume.

A TREA-

A
TREATISE

O F   T H E

EXCHEQUER and REVENUE

O F

IRELAND.

C H A P. 1.

Of the ORIGIN, JURISDICTION, and DIVISION
of the EXCHEQUER.

I T is not defigned in the following treatife to enter into
a difquifition concerning the ancient conftitution of the
Exchequer * in England; or to fhow how it was formed
from, and agreed with, that of Normandy: thofe who are
defirous of receiving information on that fubject will find
their curiofity amply gratified in the laborious and learned
refearches of Mr. Madox, in his hiftory of the Court of
Exchequer.

* The common and moft probable derivation of the Name is from the old
French word Eschequier, which fignifies a Chefs-board, or Chequer-work; and
becaufe a cloth of that kind was laid upon the table, upon which the accomptants
told out the King's money and fet forth their accounts, it was called the Court of
Exchequer. Madox 109.

Court of
Exchequer in
Ireland,
formed from
that of Eng-
land.
It is fufficient to obferve, that the Court of Exchequer
in Ireland, which is one of the four fuperior courts at
Dublin, was formed after the Exchequer in England, probably about the 12th year of the reign of King John, viz.
A. D. 1210, at which time that King caufed all the laws
and cuftoms of England to be eftablifhed, for the future,
in Ireland; as appears by a charter of Henry III. begin-
Co. Litt. 141.
b. 7 Co. 22.
b. Calv. cafe.
ning with thefe words; " Rex, &c. Baronibus, militibus,
" & omnibus libere tenentibus L. falutem. Satis, ut cre-
" dimus, veftra audivit difcretio, quod quando bonæ me-
" moriæ Johannes, quondam rex Angliæ, pater nofter, venit
" in Hiberniam, ipfe duxit fecum viros difcretos & legis
" peritos, quorum communi concilio, & ad inftantiam Hi-
" bernienfium, ftatuit & precepit leges Anglicanas in Hiber-
" nia, ita quod leges cafdem, in fcripturas redactas, reliquit
" fub figillo fuo ad fcaccarium Dublin."

and agrees
with it in its
bufinefs, &c.
And as the Exchequer in Ireland was formed from that of
England, fo it agrees with it pretty much, as well in the
names and duty of its officers, as in its bufinefs and practice; being, like that, inftituted to order and determine the
rights and revenues, and to recover the debts and duties
due to the crown.

Divifion.
According to the ufual divifion, this Court confifts, as
it were, of two parts; whereof the firft is called the judicial or fuperior part; and the other, the receipt or inferior
part of the Exchequer.

Judicial or
fuperior part.
The judicial or fuperior part of the Exchequer is converfant, efpecially, in the judicial hearing and deciding of
all caufes appertaining to the King's coffers; and was anciently called, Scaccarium computorum. And this part of
the Exchequer is a Court of law and equity.

The

The Court of law, or plea fide, is held, after the courfe of the common law, before the Barons. And here the plaintiff ought to be farmer or debtor to the King, or fome way accomptant to him. And in this Court the Attorney general brings his information for any matter touching the King's revenue. And the leading procefs is either à writ of Subpena, or *quo minus.* A Court of law.

The Court of equity is held before the Treafurer, Chancellor, and Barons; but ufually before the Barons only. The proceedings are by Englifh bill, and in a great meafure agreeable to the practice of the High Court of Chancery. And the plaintiff muft here likewife fet forth that he is debtor or farmer to the King. In this Court the clergy ufually exhibit bills for the recovery of their tythes. And here the Attorney general brings bills for any matters concerning the revenue. And any perfon grieved in any caufe profecuted againft him in behalf of the King, may bring his bill againft the Attorney general to be relieved in equity. And a Court of Equity.

And by a ftanding rule in the equity fide of the Court upon filing any bill againft the Attorney general to be relieved againft any information, fcire facias, or other matter, he fhall not be ferved with a fubpena to anfwer, but fhall be attended with an attefted copy of the bill, and an order defiring him to anfwer in four days after fuch fervice; which order the Chief remembrancer is to enter of courfe. And if the Attorney general fhall fail to anfwer within that time, upon affidavit made of fuch fervice and motion thereupon, an order fhall be granted to ftay proceedings, until anfwer or further order of the Court. And his anfwer is ufually fworn to. Rule as to proceeding againft the Attorney gen.

And

And the Attorney general may call upon any that are interested in the caufe, or any officer or others, to inftruct him in the making of his anfwer, fo as that the King be not prejudiced thereby.

But now, by a fiction, all kinds of perfonal actions may be profecuted by any perfon in this court. For as all the officers and minifters of this court have, like thofe of other fuperior courts, the privilege of fuing and being fued only in their own courts, fo alfo the King's debtors and farmers, and all accomptants of the Exchequer, are privileged to fue and implead all manner of perfons in the fame court that they themfelves are called into. So that by the fuggeftion of privilege, any perfon may be admitted to fue in the Exchequer as well as the King's accomptant; and the furmife of being the King's debtor is become mere matter of form and not traverfable. And the fame holds with regard to the equity fide of the court; for there any perfon may file a bill againft another upon a bare fuggeftion that he is the King's accomptant.

Every action, which concerns the King's revenue immediately, muft be fued in this court; and if brought in another court will be removed hither. As where an ejectment is brought by a perfon, whofe title is under an extent out of this court, for debts in aid. So if a man be outlawed in a civil action, and lands in his poffeffion be extended, and a third perfon who claims a title to them brings his action, it muft be in this court.

So where an extent in aid was taken out by the King's farmer of the hearth-money againft his own debtor,

againft

against whom a commiffion of bankruptcy was before
awarded, and the affignees under the commiffion brought
their bill in Chancery to fet afide the extent in aid; the
bill was difmiffed, for that the court of Chancery had no
jurifdiction in cafes of this nature, which were only
proper for the court of Exchequer, from which the extent
iffued, and where it was examinable.

So the officers of the revenue ought to be fued in this
court for what they do in the execution of their office,
and the court will remove an action, commenced in another court, againft an officer, for feizure of a fhip, though
no information for the fhip be yet filed. Dunb. 34.

So if trover be brought in another court againft a
cuftom-houfe officer, for tea and other goods feized by
him, and condemned, and other articles are thrown into
the declaration, to give colour to the action there; the
court of Exchequer will remove the action. Bunb. 309.

But where an officer of cuftoms feized two cables, one
of which was condemned and forfeited, and the owner
brought trefpafs in B. R. againft the officer for taking a
large quantity of cordage generally, it not appearing but
that the action was brought for that cable only which was
not condemned, the court of Exchequer would not
remove the action. Bunb. 306.

And where a perfon was fined and imprifoned by the
Commiffioners of Excife in England, and brought his
action for falfe imprifonment in B. R. the court of Exchequer would not remove the action, becaufe it did not
immediately concern the revenue of excife, but was a
penalty impofed for an offence committed in it; and fo
belonged Hard. 193.

belonged no more to the court than other like cafes arifing upon fines and imprifonments.

Jurifdiction
of the court
as to receiv-
ing pleas of
difcharge.
Somers arg.
22. It appears that formerly the jurifdiction of this court was very defective, in what feemed abfolutely requifite for the doing juftice to the fubject, who, upon accounting, was put to his petition for a writ or letter of the great or privy feal, for juft and reafonable pleas by way of dif-charge. But by the 5 Ric. 2. c. 9. Eng. it is ordained, " that the Barons fhall from thenceforth have full power to hear every anfwer of every demand made in the Exche-quer; fo that every perfon impeached, by himfelf or by any perfon, fhall be received to plead, fue, and have his reafonable difcharge, without tarrying for, or fuing, any writ, or other commandment."

The different
rolls of the
court. In general the bufinefs and acts of the court of Exche-quer were anciently entered or recorded in feveral rolls; the principal whereof, befides the plea rolls, were the *Rotulis Annalis*, or great roll of the Pipe, and the *Memoranda*.

The great roll
of the pipe.
Madox. 618. Amongft the records of the Exchequer, the great roll of the pipe muft be placed firft, by reafon of its pre-eminent dignity. It was and is the moft ftately record in the Exchequer, and the great medium of charge and dif-charge of rents, ferms, and debts due to the Crown. Into it the accompts of the ancient royal revenue were entered through divers channels. And the authority of it was fo great, that when debts had been put in charge there, they could not be difcharged unlefs by judgment or award of the chief Jufticiary, or of the Treafurer, the King's Chan-cellor, or his Council or the Barons.

The

The records or bundles made up by the two Remembrancers of the Exchequer, have been ufually called *Memoranda* or the Remembrances. A Remembrance was anciently wont to be made for every year in each of the Remembrancer's Offices. In thofe *Memoranda* there was anciently entered great variety of bufinefs; for inftance, the King's writs and precepts of many kinds, relating to revenue tenures, commiffions of Bailiwicks, cuftodies, ferms, &c. prefentations and admiffions of officers of the Exchequer, &c. pleadings and allegations of parties, judgments and awards of the Court, recognizances of debts, and conventions of divers kinds, accompts and views of accompts; with feveral acts relating to accomptants; inquifitions of fheriffs, efcheators, &c. advents of fheriffs, efcheators, &c. and in general all thofe things which were comprifed under the term, *Communia*, or common Bufinefs.

The Memoranda or Remembrancers. Madox 619.

The other part of the Exchequer, called the receipt of the Exchequer, or the inferior Exchequer, or Treafury, is properly employed in the receipt and payment of money; and in England this is a diftinct court and wholly under the Treafurer. And if any orders are fit to be abolifhed in the Receipt, or any new orders to be made, it is done by the Lord Treafurer, and ufually with the concurrence of the Chancellor and under Treafurer. If the King thinks fit to command, by privy feal, that any new order or method fhould be obferved in any part of the receipts, it is ufually directed only to the Treafurer, and the Chancellor, and under Treafurer. And if it be thought proper that it fhould be publifhed and enrolled in the Court of Exchequer, to the end that all officers and accomptants might the better take notice of it, the Lord

The receipt of Exchequer or inferior Exchequer. Somer's arg. 53. &c. Madox 179.

Treafurer

Treafurer, and Chancellor and under Treafurer come into the court of Exchequer, and the Treafurer commands it to be publifhed and enrolled, together with his own affent to it, and the affent of the Chancellor and under Treafurer; but no notice is taken of the Barons in any part of the bufinefs.

The Barons have no power over it. Somers argument 5 Mod. 45, 62.

It was determined by Lord Somers, on a writ of error brought in the Exchequer-chamber, on a judgment in the Exchequer in England, in the cafe of Hornby, &c. againft the King, commonly called the bankers cafe (tho' contrary to the opinion of all the Judges, except C. J. Treby) that the Barons of the Exchequer could not, upon the prayer or petition of a grantee of any branch of the revenue to them immediately, order the Treafurer or Chamberlain to pay out of the receipts of the Exchequer the arrears or growing payments; but that fuch grantee muft refort to his petition of right ; for that their power over the King's treafure is only *in tranfitu*, and that the law has intrufted the King himfelf only with his treafure, when once it comes into his coffers.

Irifh treafurer accomptable anciently to the Exchequer in England. Madox 633.

It appears from many inftances mentioned by Madox, that the King's Treafurer in Ireland, in the earlieft times, accounted at the Exchequer of England for his receipts out of the King's treafure at the Exchequer of Dublin, by the counter-rolls of the latter Exchequer exhibited at the former.

And King Edw. I. in the 21ft year of his reign, commanded that for the future the accompts of Gafcony and Ireland, fhould be rendered yearly at the Exchequer of England, before the Treafurer and barons there; viz. the
former

former by the conftable of Bourdeaux, and the latter by the Treafurer of Ireland.

C H A P. II.

Of the SEVERAL and RESPECTIVE OFFICERS belonging to the EXCHEQUER, both SUPERIOR and INFERIOR.

THE officers of the Exchequer may be diftinguifhed into thofe of the fuperior Exchequer, and thofe of the inferior Exchequer.

Officers of the Exchequer diftinguifhed,

The officers of the fuperior Exchequer, are as follow. into thofe of the fuperior.

The LORD HIGH TREASURER.

He is the third great officer of the Crown in Ireland, and the higheft officer both of the fuperior and inferior Exchequer, and his office is as ancient as the eftablifh-ment of the Englifh government here; he was in all an-cient writs and records called, Treafurer of the Exchequer. He is the chief judge in all caufes, that are inftituted by Englifh bill, in the chancery or equity fide of the court. And by 10. H. 7. 1. it is enacted, that the Treafurer of Ireland, fhould have as ample power in all things belong-ing to his office, as the Treafurer of England; as to make cuftomers, comptrollers, farmers, and other officers, ac-comptants for the greater increafe of the King's revenue in Ireland; and that he fhould every year make a decla-ration of his accompt of the revenue before the Barons of the Exchequer, and before fuch of the King's Coun-cil there, as fhould be appointed by the King's Lieutenant

Lord high treafurer.

or deputy; the fame declaration to be certified into the Exchequer in England : and there the accompt to be determined before the Barons. But notwithflanding this act, the Lord Treafurer enjoys very few of the privileges, which belong to his office. For the Vice Treafurer has the receiving and iffuing of all the revenues, both annual and cafual; and all the faid offices are granted by the Chief Governors for the time being, by their own warrants, and not by the warrant of the Lord Treafurer.

The Chancellor of the Exchequer.

Chancellor of the Exchequer. Madex 130, 580. 4 loft. 104, 115.
In the ancient Exchequer, this was a very great officer. He was one amongft the Juftices and Barons that ufually fat there, and tranfacted feveral things in the Exchequer in fuch manner, as that we may fuppofe it to have been anciently part of his duty to affift there. He feems to have been a control or check on the Treafurer. He has the cuftody of the feal of the court, and is a Judge in matters of equity.

The Lord Chief Baron, second, and third Baron.

Barons.
Thefe have judicial power in all caufes of law, equity, and revenue. In the two firft they govern themfelves by the common methods of proceeding in the courts of Chancery, King's Bench, and Common Pleas; and in the laft by rules of their own, and in refpect hereto the court is always open, as well out of as in Term. They are called Barons of the Exchequer, becaufe in England Barons of the realm were occafionally fummoned and fat there, with other great Officers of State. Upon their entrance into office, they take an oath not to refpite or

protract

protract the King's bufinefs, but to give it preference to all others.

The Lord Chief Baron.

He is at this day the chief Judge of the court in matters Chief Baron. of law, information and pleas. Therein he anfwers the bar, and all fuitors; and gives orders for judgment thereupon. He alone fits as Juftice of Nifi Prius to try all iflues joined in this court for the city and county of Dublin; but in his abfence one or both of the other Barons may be Judge or Judges of Nifi Prius. He takes recognifance for the King's debts, for appearances, and for obferving of orders. He takes the prefentation of all the officers in court under himfelf, and of the Lord mayor of the city of Dublin.

The Auditor General.

He is an officer both of the fuperior and inferior Ex- Auditor General. chequer. In his office are entered *Verbatim* all * grants of land and offices, whereon any rent is payable to, or ftipend payable by the King; and from thence he makes out rolls of all the King's rents; with him are lodged all the deeds of affignment or purchafe of lands, &c. out of which any rent or duty is payable to the King; (otherwife the rent is always continued in the name of the former proprietor;) and he gives conftats or certificates of fuch rents when demanded. In his office are likewife lodged all the accounts of the Vice treafurer, and clerk of the

* It was determined in the cafe of the King v. Daly 12 Dec. 1747, that a book from his office, in which patents and grants of lands are entered, is evidence. But it muft be proved by a clerk of the office, to be a book belonging to the office, and brought from thence.

hanaper,

hanaper, and of money imprefted for any particular ufe or fervice; which accounts are made up by him, and paffed by the Commiffioners of accounts, after they are compared by them with the vouchers, and by the Auditor with the books returned to him by the vice treafurer, chamberlains, and clerk of the pells. He collects a particular of what rents are unaccounted for, and remain in arrear, and tranfmits the fame to the Treafurer's remembrancer, to iffue procefs for the levying thereof.

The SURVEYOR GENERAL of LANDS.

Surveyor Ge-
neral.

He has in his office all the furveys of the King's lands, &c. and if any controverfy arife concerning the extent or boundaries of them, he appoints furveyors to fettle the mears, and bounds, and the quantity of fuch eftate, together with the yearly value thereof (if required;) for which purpofe commiffions are iffued by order of the Exchequer, on which inquifitions are taken by a jury; and the efcheator fometimes affifts therein.

When a grant is to be paffed for any eftate, a warrant from the Lord Lieutenant is directed to him, and to the Auditor general, to make out a particular of the eftate; the Surveyor general makes out the furvey in parchment, and gives the Auditor the particular; and, out of the furvey, he afcertains the rent payable to the Crown; the furvey remains with the auditor, and the particular, when examined and figned, is tranfmitted to the Lord Lieutenant under both their hands; upon which a warrant is made directed to the Attorney general, to prepare a *fiat* purfuant to fuch particular.

The

The REMEMBRANCERS *(Rememoratores.)*
Formerly called Clerks of the Remembrance.

Of thofe there are two diftinguifhed by the names of
the *King's* or *Chief remembrancer*, and the *Lord treafurer's*,
or *fecond remembrancer*.

Remembran-
cer.

The KING's REMEMBRANCER, or CHIEF
REMEMBRANCER.

He is a principal officer of the court, of great truft. In his
office all bonds for the King's debts, alfo, all recognifances
taken before the Barons for any of the faid debts, for appearances, and for obferving orders, *&c.* are entered or
lodged; and he makes out the neceffary procefs thereon.
All informations upon penal ftatutes, and upon forfeitures, and efcheats, either at law or in equity, and the
pleadings, and proceedings thereon, are filed in his office.
All inquifitions upon commiffions out of this court to
find out the King's title to any lands, *&c.* forfeited or efcheated to the crown, (efpecially to thofe which were forfeited by the rebellions of 1641 and 1688,) are returned
thither: (fuch as are held upon commiffions out of the
court of Chancery being in the rolls office;) as alfo feveral of the proceedings upon the acts of fettlement and
explanation; as the certificates of the commiffioners for
executing faid acts, decrees of innocence, *&c.* and likewife all inquifitions on *levari facias*, for the King's debt,
and *cuftodiams* and injunctions, are thereupon made out by
order of the court. All englifh bills in this court, and
the pleadings and proceedings thereon remain in his office.
He has the entering of all pleas, judgments, *&c.* relating
to the King's revenue. In his office are all the books relating to the cuftoms and excife. He makes out procefs
against

King's re-
membrancer.

againſt the collectors of the cuſtoms and excife, &c. for their accounts, not being by act of parliament directed to be otherwiſe managed than according to the courſe of the court. All diſputes touching irregularities in the practice and proceedings in the court are referred to him by the court. He has in his office all the reducements and abatements of quit rents, which were made by the Lord Lieutenant or other Chief Governor and Council, purſuant to the act of explanation. He has alſo all the reducements of fines by the commiſſioner of reducements. He reads in court the oaths of all the officers, attorneys, and miniſters of the court, when they are admitted; and he alſo reads the oath of the Lord mayor of the city of Dublin, and of all the ſheriffs in the kingdom; writs of prerogative or privilege for officers and miniſters of the court are made by him. All ſums of money brought into court by order are lodged with him, although no ſecurity is required of him, on his entring into office. He has in his cuſtody the red book of the Exchequer.

He has under him five ſecondaries, a filacer, and other aſſiſtant officers. One of the ſecondaries has the office of the law pleas of this department.

The Lord Treasurer's, or Second Remembrancer.

Treaſurer's Remembrancer.

He takes notes of all rules and orders made in the court, relating to the King's revenue, except the cuſtoms, excife, and other ſuch revenues which are in the Chief remembrancer's office. He makes proceſs againſt all ſheriffs, eſcheators, receivers, and bailiffs for their accompts. He makes proceſs of *fieri facias* and extent for any debts due to the King either in the Pipe or with the Auditors. He makes a record, whereby it appears whether ſheriffs and

other

other accomptants pay their *profers* due at Eafter and Michaelmas. And he makes another record, whereby it appears whether fheriffs and other accomptants keep their days of prefixion, or days appointed. Into this office are certified all eftreats of fines, iffues, *&c.* fet in the fuperior courts at Dublin, or at the affizes or feffions. And into it are returned all the inquifitions upon the writs of *levari facias* and of feizures, which iffue thereout for the King's rents, and alfo upon the writs of *levari facias* which iffue thereout for fines, forfeited recognifances, and other matters eftreated into the office; as likewife the certificates of fheriffs as to the goods of felons and fugitives, waifs, eftrays, *&c.* in their bailiwicks. All the pleadings, orders, and proceedings touching the reducing, exonerating, refpiting, or difcharging any of the King's rents, fines, *&c.* are in this office; as are alfo the certificates of the commiffioners of reducements, which are fent thither by the Chief remembrancer, on which an order is entered here to be taken to the Clerk of the Pipe to make out the *debet*; which *debet* is to be taken to the treafury, and, the money being paid there, an acquittance is given, which is to be brought to the Second remembrancer, who thereupon enters an order for the abfolute difcharge of the fame. Formerly, tranfcripts of all the grants, that were paffed by the crown, of any lands, *&c.* were brought in twice every year by the mafter of the rolls, (as it is faid) and delivered to the court, and, by the court, tranfmitted into this office, to be compared with the entries in the Auditor's office, left any thing fhould pafs not entered by him; this office being a check to the Auditor's office as to the King's rents. In this office likewife were, formerly, enrolled all claims of privileges, franchifes, liberties, *&c.* licenfes of alienation, pardons of alienation, grants of goods of felons,

fugitives,

fugitives, and outlaws, waifs and eftrays; but this does not appear to have been practifed fince Eafter term 1686. This officer has, likewife, under him two fecondaries.

The CLERK of the PIPE.

Clerk of the Pipe.

He makes out the yearly roll, which is called the great roll of the Pipe, of all rents and debts whatfoever, that are brought in by procefs to any of the other officers of the Exchequer, and accounted for in the court; and of all the debts that are in arrear and unanfwered for by the fheriffs on paffing their accounts. He alfo writes fummonfes to the fheriffs to levy the faid debts upon the goods and chattels of the debtors, and if they have no goods, then he draws them down to the Lord Treafurer's remembrancer to write extents againft their lands. He makes a charge to all fheriffs of the fummons of the pipe and green wax, and fees that it is anfwered upon their accounts. All orders of difcharge and refpite whatfoever of any fuch rents or debts are entered with him. He makes out cuftodiams upon feizure or fequeftration of any eftate, and outlawry eftreats upon procefs from the Treafurer's remembrancer's office; and, from time to time, renews procefs for all fuch arrears as ftand out upon the roll. He has the drawing and engroffing of all leafes of the King's lands. And in this office the fheriff's *quietus eft* is prepared, as being the laft office of account of the procefs.

The COMPTROLLER of the PIPE.

Comptroller of the Pipe.

He writes out fummonfes twice a year to levy the farms and debts of the Pipe, which is called the fecond fummons, and is in the nature of a *levari* againft the body, goods, and lands of the debtor; and he alfo keeps a comptrol-

ment

ment of the pipe or counter roll of all arrears; he is affifting to the clerk of the pipe, and iffues out the fecond
procefs of the pipe, by warrant from the clerk of the pipe,
and counter-figns it with the clerk of the pipe.   And he
alfo, as well as the clerk of the pipe, takes down what is
nihilled by the fheriff on his accounts, and the fum he
charges himfelf with.

The CLERK of the ESTREATS and SUMMONISTER.

Towards the reign of Edward III. the cafual revenue
being fo much increafed, that the clerk of the pipe could
not engrofs all the fums eftreated on his annual roll, and
many of them being fmall, and paid on the firft demand,
it was neceffary to make them part of the annual charge,
in the fame manner as the other annual revenue of the
King was; therefore a new officer was created, viz. the
clerk of the eftreats; and inftead of delivering the eftreats
of the Exchequer and other courts to the clerk of the pipe,
they were, thenceforward, delivered to him, and he iffued
a diftinct procefs from the fummons of the pipe, viz. the
fummons of the green wax, which is the firft procefs; and
hence in this kingdom he is alfo called the *fummonifter*.

Clerk of the
eftreats and
fummonifter.

And this officer, as well as the clerk of the pipe, receives
the anfwer of the fheriff in court; and the *nihils* are to
be entered on the great roll.

As clerk of the eftreats, he has the care of all fines,
amerciaments, and cafualties, that arife in any of the
courts of record; and of the fines and amerciaments that
are impofed in the Exchequer, in the King's remembrancers office and pleas office, or at the affizes or feffions,
which are brought into the clerk of the eftreats, by the

VOL. I.   D   refpective

rcfpective officers of the courts, and clerks of the peace ; upon which he makes out procefs; which is tranfmitted under the feal of the court to the fheriffs, to levy fuch fines, forfeitures, and debts; and this procefs is in the nature of a *fcire facias*, which notwithftanding the fheriff muft anfwer in his accounts, or take bonds from the party, to clear fuch debts in court.

The Transcriptor and Foreign Apposer *.

Tranfcriptor and foreign appofer.

He is an officer in the Exchequer, to whom all fheriffs and bailiffs repair, to be appofed by him of their green wax, after they are appofed of their fums out of the pipe office; and from thence he draws down a charge upon them to the clerk of the pipe. His bufinefs is to examine the fheriff's eftreats with the record, and to afk the fheriff what he fays to every particular fum therein; he fends the debts *nihilled* by the fheriff to the clerk of the pipe, which then being prefumed to be debts that will ftand out for fome time, are by him entered on the great roll, and *debets* thereof fent by him twice a year to the comptroller of the pipe; who fends out the fecond procefs of the pipe ; becaufe having already been in procefs on the fummons of the green wax, it hath anfwered *magna charta*, 9 Hen. 3 c. 8. by which no fheriff or bailiff fhall feize lands for the King's debts, fo long as the prefent

* In Gilbert's treatife of the Exchequer, this officer is faid to be called *appofer*, for the fame reafon that the fheriffs accounts of their green wax were called *appofals*, viz. becaufe the fheriff was then *apponere*, or to *place* his items to account. And he is called the *foreign appofer*, becaufe the account on which he fat, was a *foreign* and diftant account from that of the great roll, which was carried on by itfelf; or becaufe this cafual revenue, not arifing out of originals fent into the court from the Chancery, as the certain revenue did, which was the *original* jurifdiction of the court, but being fent into the Exchequer by eftreats out of other courts, was therefore called the *foreign* revenue. See Madox 708.

goods

goods and chattels of the debtor shall suffice, and the debtor be ready to satisfy the same.

The CLERK of the PLEAS.

In his office are all the proceedings at law between party and party, under the surmise or fiction of the plaintiff's being the King's debtor, and not immediately concerning the revenue. *Clerk of the pleas.*

The SERJEANT at ARMS.

He is an officer attending this court, as likewise the house of Commons. To him is directed the last process of contempt, on which a sequestration is grounded. *Serjeant at arms.*

The PURSUIVANT.

He was anciently a messenger attending the King in his wars, or at the council table, or in the Exchequer, to be dispatched upon any occasion or message. All sheriffs and coroners, and all officers of the court (except the marshal and usher) for misdemeanors, mis-execution, or non-execution of their office, and all persons guilty of any special contempt in this court, are committed to his custody. *Pursuivant.*

The USHER.

It was his duty to keep the Exchequer safely, and to take care of the doors and avenues of it; so as that the King's records which were laid up there might be in safety. It was also his duty to transmit the writ of summons which issued out of the Exchequer for the King's debts; that is, to cause them to be delivered to the *Usher. Madox 727.*

several

feveral fheriffs to whom they were directed. It is likewife his duty at prefent to furnifh the court with books,
paper, and fuch neceffaries.

The MARSHAL.

Marfhal.
4 Inft. 107.
Madox 727.

To this officer the court committed the cuftody of the
King's debtors during the fitting of the term, to the end
that they might provide to pay the King's debts, or elfe
be further imprifoned. Such offices as were found *virtute*
officii, and brought into the Exchequer, were delivered to
him to be delivered over to the treafurer's remembrancer.
He alfo appointed auditors to fheriffs, efcheators, cuftomers and collectors, for taking their accounts.

But this office and that of the ufher are now exercifed
by the fame perfon; and yet in the patent to the ufher
there is no mention made of the marfhal.

To the officers of the fuperior Exchequer may be added,
as being attendant thereon, the King's Attorney general
and Solicitor general.

The ATTORNEY GENERAL.

Attorney ge-
neral.

He has a fpecial charge of the revenue. No rent can be
difcharged or abated, but by his conceffion or confent.
He puts into court, in his own name, informations of con-
cealment of cuftoms, feizures, &c. alfo of intrufions,
wafte, and encroachments upon any of the King's lands,
and upon penal ftatutes, forfeitures, &c. He acts in
general by *debet*, *conflat*, or certificate, from the proper
officers, in whofe office the debt or matter, for which the
information or bill is brought, is recorded. He prepares
the

the *fiats* for all patents for lands, eftates, honours, &c.
He is made privy to all manner of pleas that are not ordinary, and of courfe, which arife upon the procefs of the
court; and he is an officer of fuch dignity and confidence,
that his confeffion binds the crown in all fuits and caufes
wherein he is concerned.

The Solicitor General.

He is affifting to the Attorney general in all the matters
aforefaid, as to pleadings in court; and he, as well as the
Attorney general, may prepare the *fiats* for patents.

Solicitor general.

The officers of the Inferior Exchequer, called the receipt of the Exchequer, are as follow:

The Lord Treasurer,

Of whom mention has been already made amongft the
officers of the Superior Exchequer, being the Chief
officer both of the fuperior and inferior Exchequer, and of
whom, as he now feldom acts, the whole bufinefs being
done by the Vice treafurer or his deputy, there is no occafion for adding any thing more.

Lord treafurer.

The Vice Treasurer.

He is a principal officer of this part of the court, under
the feveral appellations of Vice treasurer, Receiver
general, Pay-master general, and Treasurer at
War.

Vice treafurer.

He has the charge of all his Majefty's revenue, of what
nature or kind foever, and is to account for the fame. He

Has the charge of the revenue.

or

or his deputy is to fign all receipts for money paid into the treafury, and received for him by the teller or cafhier; as alfo all orders and debentures for money paid out. And no acquittance that is not figned by the Vice treafurer or his deputy (except thofe given by the fheriffs for debts iffued in green wax) is or can be a difcharge to the fub-ject for any fum paid in, &c. except where fome act of parliament, or the King, by letters patent under his great feal, fhall otherwife appoint.

Pays the civil and military lift. As Vice treafurer and Receiver he only pays the civil lift; and what remains, after the payment thereof, is tranf-ferred by him to the difcharge of the military lift, which he pays as Treafurer at war and pay-mafter.

The civil lift by deben-tures. The civil lift is paid by debentures made out by the Auditor general, and figned by him purfuant to the King's eftablifhment, a copy of which is lodged with the Vice treafurer.

The military lift by war-rants by the Commiffary general. The military lift is paid by warrants prepared by the Commiffary general of the mufters, and figned by the Lord Lieutenant or other Chief Governors, and counter-figned by the Commiffary general, and not by the fecretary at war, as is ufual in England.

Payments on the head of Concordatum how made. Payments on the head of the Concordatum * in the civil lift, are to be on warrants, to be moved for and granted at the council board, and thefe warrants are to be

* This is an annual fum of £5000, limited in the civil eftablifhment, to be paid as fhall be *agreed*, (from whence it has its name) that is to fay, by *Concordatums* of the Lord Lieutenant, Lord Deputy, Lords Juftices, or other Chief Governor or Governors and Council.

And

be figned by the Lord Lieutenant, or by the other Chief Governors, and a *quorum* of the council; and if the Vice treafurer exceeds, in his payments or warrants of *Concordatum*, the fum limited in the eftablifhment, he is liable to refund and to make all fuch overplus payments good to the King.

And the general authority for all payments out of the treafury of this Kingdom, is the eftablifhment of the civil and military lift; which is figned by the King, and Lords of the treafury, and tranfmitted over hither. And the Vice treafurer cannot pay any warrant that comes to his hands, figned by the King, or any warrant of privy feal, or warrant under the great feal of England, for any fum or fums of money not included in the eftablifhment, without being liable to have the fame chequed and difowned

Civil and military lifts, his general authority for all payments out of the treafury.

And this fum is, in the eftablifhment, expreffed to be for freight, tranfportation, carrying of letters, and other expenfes and rewards; fea fervice, repairing and upholding fufficiently, the King's houfes; maintaining his forts; finifhing needful undertakings of that kind, begun in apt places but not finifhed; erecting of more ftrength of the like nature in other fit and neceffary places; diet and charge in keeping poor prifoners, and fick and maimed foldiers in hofpitals; printing, riding, and travelling charges; prefts upon account, and all other payments; amongft which, the repairs of fortifications, and provifion of hofpitals are chiefly to be taken care of; and thefe *Concordatums* are to be every three months certified over to the privy council in England.

And no payment or allowance is to be made by *Concordatums*, but by warrant drawn by the clerk of the council in Ireland, paffed openly at the council board there, and figned by the Lieutenant, Deputy, or other chief Governor or Governors, and by three or more of the officers following, *viz.* the Chancellor, Treafurer, Vice treafurer, and Chancellor of the Exchequer, the two Chief Juftices, Chief Baron, Mafter of the rolls, and Secretary of ftate; and for default, either by exceeding the fums limited, by anticipation, or by not obferving the faid direction in every point, all the fums that fhall be otherwife allowed and paid there, fhall be fet *infuper*, as debts upon the Lieutenant or any other Chief Governor or Governors of Ireland, the under Treafurer, and all others that fhall fign the fame, to be defaulked, to the ufe of his Majefty, upon their feveral entertainments.

upon

upon his accounts; unlefs the payment he makes on fuch
warrant be alfo purfuant to directions in writing from the
Lord Lieutenant, or the other Chief Governors, and a
quorum of the council thereupon.

Not to pay debentures or warrants without entering them with the clerk of the pells.

He is not to pay any debentures or warrants on the civil
or military lift, or any other account, without entering
them with the clerk of the pells, and having them counter-figned by him; otherwife fuch payments will be difallowed in his accounts; fo that he can neither receive or
pay without a control; by means whereof the King cannot be defrauded.

To give Exchequer acquittances for all money paid to him.

For all money paid into the treafury, he or his deputy,
is to give an acquittance, which is to be figned and entered
with the Clerk of the pells, Chamberlains, and Accountant
general, and to be delivered to the party paying his
money. And this is called an Exchequer acquittance *.
And when thefe acquittances are returned by the Accountant general or other perfon, as vouchers, they are to be
filed of record with the Auditor general, where they are
to remain, and are a charge on the Vice treafurer.

His accounts to be taken by Commiffioners.

His accounts are not to be taken by the Barons of the
Exchequer, but by Commiffioners authorized under the
great feal of England or Ireland. And upon his accounting before the Auditor and Commiffioners, he delivers in
books containing tranfcripts of every individual receipt,
and fum of money by him received; and the clerk of
the pells doth the like; which books remain with the

---

* Formerly, a lift of thefe acquittances was brought weekly from the treafury
to the Accountant general ; but this caufing great confufion in the accounts, as to
dates, &c. the prefent method is purfued of fending them to the Accountant
general to be entered as they are paffed.

Auditor

Auditor general, as a difcharge both for the Vice trea-furer and the fubject who paid his money. He alfo gives the Auditor general a fair tranfcript of all his payments under the feveral heads of the eflablifhment, expreffing at large the nature of the payments, and for what time &c. and when the Auditor general has examined and engroffed his account in parchment, it is brought with the vouchers to the Commiffioners of accompts, who fit, and examine and compare the fame; and they being fa-tisfied therewith fign it.

When his account is thus paffed, his *quietus cft* is a his quietus. duplicate thereof figned by the faid Commiffioners, which he keeps; and the other remains with the Auditor ge-neral.

The TELLER or CASHIER.

He receives all the King's money, and afterwards Teller or Ca-writes a bill in parchment for the party's acquittance who fhier. pays it; which acquittance is tranfmitted to the Cham-berlains, who enter and fign it; and it is then delivered by them to the Clerk of the pells, to be entered and fign-ed by him; and is then delivered to the party.

He alfo pays out the King's money upon debentures, and by orders from the Lord treafurer, and under trea-furer, which are directed by the auditor.

The CLERK of the PELLS.

He enters all the teller's bills into a book or parch- Clerk of the ment roll, called *pellis receptorum.* He figns, and enters Pells. all acquittances that are given by the Vice treafurer, and VOL. I.   E   Receiver

Receiver general or his deputy, for any rent, debt, or sum of money paid into the treasury. With him are also entered all warrants of debentures, upon which the Vice treasurer or Receiver general makes any payment; and hereby he is a perfect cheque upon the Vice treasurer, and ought to be able at any time to give a state of his receipts and payments, and to know what money he has in his hands to answer the King's affairs. He returns into the auditors office, yearly, books containing every individual acquittance, passed by the Vice treasurer, and Receiver general, or his deputy for that time; which the auditor compares with the like books returned to him by the Vice treasurer and chamberlains, that so the King may not be prejudiced, but the Vice treasurer fully charged with all the money received by him during the time for which he accounts.

The CHIEF CHAMBERLAIN and SECOND
CHAMBERLAIN.

Chamber-
lains.
Madox 732,
Somers arg.
51.

They are ancient and were great officers of the superior Exchequer in England; and sometimes sat and acted in person, and were numbered with the Barons there. But the office being such as might be executed by deputies the chamberlains by degrees made themselves useless, by leaving all the business to their deputies, and the office itself sunk by degrees to little more than a name. Their business, at this day, is pretty much the same with the Clerk of the pells, as to all receipts and payments into the treasury; and their books, as well as those of the Clerk of the pells, are transmitted yearly to the Auditor, not only to adjust the charge of the Vice treasurer and Receiver general, but to be compared with rent rolls in the auditor's office, whereby he ascertains

what

what rents are received, and what remain in arrears, in order to the iffuing out procefs for fuch as remain unpaid.

The AUDITOR of FOREIGN ACCOUNTS and IMPRESTS.

He audits all accounts of money impreſted for the buy- *Auditor of foreign accounts and impreſt. Madox 727.* ing of arms, ammunition, and proviſions, and all money iſſued by impreſt for the building of fortifications.

He is alfo affiftant to the Commiffioners for the impreft accounts, for the examining and making up all foreign or martial accounts or imprefts within the kingdom of Ireland, except the Treafurer's accounts for the wars. To thofe feveral officers may be added,

The COMMISSIONERS for the TREASURY ACCOUNTS.

They are the lord high Chancellor, the Chancellor of *Commiſſioners for the publick accounts.* the Exchequer, the lord chief Baron, and the Barons of the Exchequer for the time being, who are authorized under the great feal of England or Ireland; and before any three or more of thefe commiffioners, the Vice treafurer and Receiver general is to account, formerly but once, but now four times in every year; and they are to examine his accounts, and the vouchers; and to compare them; and, being fatisfied therewith, are to fign them: but before his accounts are fully cleared and difcharged, they are further fubjeft to the examination of the treafurer or commiffioners of the treafury of Great Britain.

Thefe Commiffioners had likewife formerly under their *Other accounts formerly under their infpec- tion but now* infpection feveral other accounts, fuch as thofe of the ordnance, the board of works, money advanced by govern-

E 2   ment,

under the
Commiffion-
ers of foreign
accounts.

ment by way of imprefts, &c. But they are now confined merely to the treafury accounts; and all thefe other
accounts are made fubject to the examination of five
Commiffioners, conftituted for that purpofe, called the
Commiffioners of imprefts and foreign accounts. But
thefe are not properly to be confidered as officers of the
treafury.

C H A P.   III.

OF THE PUBLICK REVENUE OF IRELAND.

AS a preliminary introduction to the prefent ftate of
the revenue of Ireland, it might have been a matter
of curiofity to have entered into an hiftory or detail of the
ancient revenues of the crown, the feveral branches
which compofed them, and the manner of levying,
accounting for, and paying them into the Exchequer. But
this fubject I fhall leave to the antiquarian, who has
abilities and leifure to make the neceffary refearches into
the publick offices and ancient records of this kingdom;
and fhall content myfelf with deducing an account of the
Irifh publick revenue, from the reftoration, (foon after
which the act of tunnage and poundage and the act of
excife were made) to the prefent time.

And this publick revenue may now be confidered as
divided into, I. The King's HEREDITARY REVENUE.
II. The ADDITIONAL DUTIES granted for the better fupport of government: And III. the APPROPRIATED DUTIES.

Hereditary
revenue.

And I. the Hereditary revenue, fo called from its being
vefted in the King, his heirs and fucceffors, and which
amounts

amounts in grofs, at a *medium* of the laft 12 years preceding 25th March 1773, to about £640,000 a year, is that which either is the ancient patrimony of the crown; or elfe was granted to King Charles II. by parliament, by way of purchafe or exchange for fuch branches of the King's inherent hereditary revenue as were found inconvenient or burdenfome to the fubject; or perhaps in lieu of forfeitures which the crown was entitled to; the produce of all which belongs to the crown, to be applied, under the conftitutional truft, for publick fervices.

Thefe may be confidered in the following order, viz. The King's rents, cuftoms outwards and inwards, import excife, prizage on wines, lighthoufe duties, ale, wine, and ftrong water, licences, feizures and forfeitures, hearth money; and the cafual revenue, confifting of fines, forfeited recognizances, cuftodiam rents, together with fome other cafualties, as waifs, eftrays, goods of felons and fugitives, &c.

Of what it confifts.

The act of re-affumption, viz. 11 Wil. III. Eng. makes the crown rents, quit rents, and chiefrys, unalienable; and enacts that they fhall for ever be and remain for the fupport and maintenance of the government of this kingdom. The act of 14 and 15 Car. II. c. 18, granting the revenue of ale licences, reftrains the crown from farming it or charging it with gift, grant, or penfion. And the act of 14 and 15 Car. II. c. 17. granting the hearth money, reftrains the crown from· particularly charging it with grant or penfion. And thefe feem to be the only branches of the hereditary revenue which the crown is reftrained from charging or aliening. See 5 Mod. 46, 54, &c.

How far chargeable and alienable.

II. The

Additional
duties.

II. The next head to be confidered are the Additional Duties, which are granted, in aid of the hereditary revenue, for the fupport of his majefty's government; and are always granted for two years certain, beginning and ending on the 25th of December; and fo far as they are granted without fpecial appropriation, they are granted to the crown under the fame conftitutional truft with the hereditary revenue.

Appropriated
duties.

III. The laft head of the publick revenue are the appropriated duties; which are impofed for certain particular purpofes, to which they are fpecially applied by parliament at the time of granting them; and thofe appropriations, at prefent fubfifting, are, the loan, the tillage, the linen manufacture, the Dundalk cambrick manufacture, the proteftant charter fchools, and the Lagan navigation.

Paid as the
others into
the treafury.

And thefe duties are paid into the treafury as all the others are; but it is only for convenience; they are feparately accounted for, and iffued by different warrants; being paid, according to the directions of the feveral acts of parliament, to the orders, or on the receipts of the corporations, or private perfons refpectively interefted therein, without any warrant figned by the government.

CHAP.

C H A P. IV.

Of the KING's RENTS in IRELAND, and the
ANCIENT and PRESENT METHOD of
COLLECTING THEM.

THERE are four feveral forts of rents in Ireland
referved and payable to the King, to wit, CROWN
RENTS, PORT-CORN RENTS, COMPOSITION RENTS, and
QUIT RENTS.

Rents payable to the King, different kinds of.

The Crown rents are ancient rents referved upon grants
made by the Crown of their demefne lands, and lands of
inheritance.

Crown rents, what.

And the greater part of thefe rents, at this day, arife
upon grants made of the lands, tenements, hereditaments,
&c. which formerly belonged to monafteries, abbeys, prio-
ries, and other religious houfes, which, in the reign of
King Henry the VIIIth, were either diffolved, fuppreffed,
renounced, relinquifhed, or furrendered to his Majefty;
and which, together with the fcites, ambits, circuits, and
precincts thereof, and all the lands, tenements, hereredita-
ments and appurtenances thereunto belonging, were afterwards, by two feveral acts of parliament, 28 Hen. 8. cap.
16. and 33 Hen. 8. cap. 5. given to, vefted in, and adjudged to be in the very actual and real feifin and poffeffion of his Majefty, his heirs and fucceffors for ever; in
as large and ample manner and form as the then late ab-
bots, priors, commanders, and other governors of the
faid religious houfes had held and enjoyed the fame; together

Out of what they arofe.

gether with all and every the rents, fervices, and rent-feck, and all other fervices and fuits which were due, to be paid, or done to any perfon or perfons from, or out of the pre-miffes, or any part thereof.

Other Crown rents.   And the rents referved on all grants from the Crown, of fairs, markets, ferries, and fifheries, are called alfo Crown rents.

Rents on grants in pur-fuance of the commiffions of grace called Crown rents.   The rents referved on the grants of the fix efcheated counties * in the province of Ulfter, are alfo called Crown rents, and are entered as fuch in the King's rent rolls.

---

* There were fix counties efcheated or forfeited to the Crown, on the rebellion of the earl of Tyrone and others, to wit, Donegal, Tyrone, Derry, Fermanagh, Cavan, and Armagh. Thefe fix counties were planted with great judgment by King James the Firft, on a plan formed by Lord Chancellor Bacon, but much improved by Sir Arthur Chichefter, afterwards Baron of Belfaft, and Lord Deputy of Ireland, and who might be faid to be the real proprietor. On this plan the lands which were to be affigned for planting, were to be fo affigned either to the old chieftains, or inhabitants, or fervitors of the Crown, (who were the great officers of ftate, or captains and officers in the army) or elfe to Englifh and Scotch undertakers; and different allotments were made to each of thefe, and encourage-ment given to them all, but efpecially to the firft, to gain their good-will.

The lands to be planted were divided into three proportions, the greateft, of 2000 Englifh acres, the middle, of 1500, and the leaft, of 1000; each and every county was fet out into thefe proportions; the one half of it affigned to the fmalleft, and the other half divided between the other two proportions.

And thefe eftates were granted by the King to thefe feveral perfons to be held by them and their heirs; the undertakers of 2000 acres held of him in capite; thofe of 1500 by knights fervice; and thofe of 1000 in common foccage. See Carte's life of the Duke of Ormond, vol. 1. pag. 14, 15, 16; as to what obligations each of thefe were under as to building, planting, with freeholders, &c.

On thefe donations of lands the following rents were referved to the Crown, viz. upon every 1000 acres (after three years exemption, and three years at half rent) a rent of 5l. 6s. 8d. from the undertakers and fuch fervitors as planted with Britifh tenants; of 8l. from fervitors that planted with the Irifh; and of 10l. 13s. 4d. from the natives, who were not obliged to build caftles.

And

And the yearly amount of the faid Crown rents is about £14800 a year.

The Port-corn rent was a kind of rent formerly paid by many of the tenants to the monafteries and abbeys before their diffolution, as aforefaid, by fervice, or in kind, by *port-corn*, or marts, or by rendering of corn, and other produce of the lands. And it is called *port*, from *porto* to carry, or, *quia ad portam monafterii jacebatur.* And the fpecies of corn fo referved, are wheat, bere, malt, and oatmeal; but in one grant beeves are referved.

And in feveral of the grants fo made by the Crown, after the diffolution of the faid religious houfes, and efpecially of rectories and tythes, and other the fpiritual poffeffions thereof, this port-corn rent has been referved, as well as the Crown rent, or rent in money.

And all thefe port-corn rents, which amount to about £400 a year, were, fhortly after the diffolution of the abbeys, &c. given by the Crown to the Lord Lieutenant, and to certain other great officers in Ireland, to wit, the Mafter of the rolls, the Lord Chief Juftice and the Lord Chief Baron, and the Prefidents of Munfter and Conaught; and they were accordingly put upon the eftablifhment for the fame, in the 42d year of the reign of queen Elizabeth; as appears by the rolls in the office of his Majefty's Auditor general, (where thofe allotted to the Lord Lieutenant are under the title, *fword)* and are faved and confirmed to them by the act of fettlement in the following words;

" Provided that neither this act, nor any thing therein " contained, fhall extend to the difpofing or altering of any " impropriate

the Lord Lieutenant or other Chief Governor, &c. in right of their places.

" impropriate rectories, or tythes, or rents, now, or lately
" enjoyed or poffeffed by, or fettled on, the Lord Lieute-
" nant or other Chief Governor or Governors of this
" kingdom for the time being, or which at any time have
" been, or are now enjoyed, poffeffed, or received by the
" Lords Prefidents of Munfter and Conaught in right of
" their refpective places, any thing in this act to the con-
" trary in any wife notwithftanding.

" And that the Lord Chief Juftice of his Majefty's court
" of King's bench, the Lord Chief Baron of his Majefty's
" court of Exchequer, and the Mafter of the rolls, or any
" other of his Majefty's officers of this kingdom for the
" time being, fhall and may have and receive fuch port-
" corn of the feveral rectories which formerly have been
" paid and referved."

To be ren-dered by the grants.

And by the aforefaid grants, the faid port-corn is to be
rendered at the principal town in the county named in the
patent, on or before the 2d day of February in every year,
to fuch perfon or perfons as the Lord Lieutenant or other
Chief Governor or Governors of the faid kingdom of Ireland for the time being, fhall, from time to time, appoint
to receive the fame.

And what the incum-bent or other proprietor may, on ren-dering it, de-falk or retain.

But in the faid grants there is ufually a claufe or provifo,
empowering the incumbent, or other proprietors of the fpiritual poffeffions thereby granted, on their producing a
bill to the Vice treafurer, or Receiver general for the time
being, teftifying the delivery of the grain, &c. to defalk
or retain in their hands, for the faid grain, two fhillings a
peck (*modius*), lawful money of Ireland. And by conflats
from the rent rolls of queen Elizabeth in the Auditor general's office, the number of pecks charged on each denomination,

mination, are mentioned; and they are valued at a certain fum for each peck, as if fuch fum was to be received in lieu of the port-corn; and the fum at which the faid pecks, payable out of each denomination, are valued, is the fame fum, which, by the faid grants, was to be fo defalked for the port-corn, delivered for fuch denomination, out of the Crown rent thereof.

The faid ftate officers being fo entitled, they ufually farmed out the faid port-corn to certain farmers, at a certain yearly rent, faid to be £200.

But fuch farmers having ufed great feverities and ex- Severities ufed by the farmers thereof. actions to the incumbents and proprietors of the fpiritual livings, a doubt arofe, when the faid port-corn had not been delivered in kind at the times and places appointed by the grants, but had been fatisfied in money or other-wife by agreement with the Chief Governors for the time being, as was ufually the cafe, whether the incumbents or proprietors of the faid fpiritual livings were entitled to de-falk any fum in their hands by virtue of the faid grants, although they fhould bring a bill or note from the perfons entitled to the faid port-corn, or the farmers thereof, to the Vice treafurer or Receiver general, if in truth the faid corn had not been actually delivered according to the grants, but fatisfaction by agreement had been made for the fame.

And complaints having been made of the faid exacti-ons and feverities, in order to remove all fuch com-plaints and grievances, it was mutually agreed upon,

and

and a * deed executed in purfuance thereof, bearing date
the 7th day of March 1698-9, between their excellencies
the Lord Duke of Bolton, and the Earl of Galway, the
then Lords Juftices of this kingdom for themfelves, and
as far as they could, for the Chief Governors that fhould
be thereafter on the one part, and by Dr. John Bolton
for himfelf and the other incumbents or proprietors of
the faid rectories, (he being by letter of attorney lawfully empowered fo to do,) on the other part, that for the
year ending the fecond of February then laft, the faid
incumbents or proprietors fhould pay for the faid port-
corn, to the faid Lords Juftices, the fum of 5s. 6d. for every
modius grani referved by the faid patent, amounting in the
whole to 269l. 18s. 3d. as in the fchedule thereunto annexed is expreffed ; and for every year after, at or before
the fecond of February in each year, five fhillings per
peck, which amounted unto 245l. 7s. 6d. per. annum:
And it was alfo further agreed, on behalf of the feveral
then incumbents, or that fhould be thereafter, and the
proprietors of the faid rectories, that none of them fhould
afk, demand, or defalk any allowance on account of the
faid corn, or by virtue or colour of their faid feveral patents, but fhould pay the feveral yearly rents payable to
the crown, as aforefaid, out of the faid rectories, without
any defalcation or deduction whatfoever, faving never-

* The rolls office, Auditor general's office, and council chamber, have been
fearched for this agreement, but it is not be found; but it is entered in a book
in the Auditor general's office. But in the year 1761, thefe matters having been
laid before the Attorney general, and others of his Majefty's council, they were
of opinion that the Lord Lieutenant, or his farmers cannot fue for the rents on
the agreement of 1698; but that the proceedings muft be in purfuance of the
grants, and for the port-corn; and that, tho' the port-corn payable by the proprietors, is by the eftablifhment affigned to feveral great officers, yet that it ftill
remained payable to the King, and that the procefs muft iffue in his Majefty's
name, for the recovery of any arrears that may be due thereon.

thelefs

thelefs, all their right and title to the faid allowance of eighteen pence per peck, as in their refpective grants and patents are referved and expreffed, in cafe the faid agreement fhould at any time thereafter be broken or made void.

And this port-corn may be recovered as others his ma- How recover able. jefty's rents, (which fee hereafter,) to wit by diftrefs, feizure, or information. And, in all the grants thereof there is a claufe, that if the fame fhall be in arrear, the King, his heirs and fucceffors, may re-enter and take the iffues and profits thereof, to his and their own ufe, until the faid arrear fhall be fully fatisfied; and then, and not before, the tenants of the aforefaid fpiritual tenures to be reftored to the poffeffion thereof.

But by the King's letter bearing date the 20th of April Now paid to the commiffioners of the revenue. 1763, it was directed that the rents ufually accruing to the Chief Governor or Governors of Ireland for the port-corn throughout Ireland, fhould be no longer paid to the faid Chief Governor or Governors for their ufe, but to the Commiffioners of his Majefty's revenue from time to time, for the ufe of his Majefty, his heirs and fucceffors. And by the government's warrant dated the 1ft. of July, in the fame year, his Majefty's Auditor general was required to make out one or more particular or particulars of the faid rents, and to put the fame in charge upon his Majefty's rent rolls, and to make a return thereof to the faid commiffioners in order to be collected by their officers, in like manner as his Majefty's other rents are collected.

And the Auditor general thereupon returned feveral conftats of charges of port-corn, which appeared to have been

been taken from a rent-roll thereof, made in the reign of Queen Elizabeth; which rent-roll is now the only evidence of such rent, where the denominations have not been granted away by the crown.

Composition rents, what

The Composition rents are certain rents reserved to the crown upon a composition made in the reign of Queen Elizabeth, between her majesty and the lords and chief-tains of Conaught, in lieu of cesses, imprefs, and quar-terage of soldiers.

And how they arose.

And the original of these composition rents was in this manner. Several lands in the province of Conaught and Munster, and other countries in this kingdom, formerly held by Irish custom, and not by tenure, according to English laws, were charged with heavy cesses and taxes, and subject to the depredations of men of war ; wherefore, at the first quieting and settling those parts under the English government, the lords and chieftains of the said provinces and countries petitioned her Majesty, by her then Lord deputy, to accept from them the surrender of all their lordships, manors, lands, tenements, and other their possessions, to the end it might please her Highness, after the said surrender so made, to grant to them the same their lands and possessions, to hold of her Highness, her heirs and successors, by such tenures, rents, services, and attendance as should be thought meet and conve-nient, respecting the quantity and quality of the said lands, &c.

Act to enable the crown to make grants of the lands accordingly.

And accordingly an act of parliament was made, in the 12th year of the reign of her said majesty Queen Elizabeth, by which it was enacted, that patents should be made out to such persons, as should surrender to the Crown

Crown their lands, fo held by Irifh cuftom, to be holden of her Majefty, her heirs and fucceffors, for fuch eftate, and by fuch tenure, rents, and fervices, as fhould be expreffed and referved in the faid letters patent.

And afterwards in the 27th year of the faid reign, a commiffion iffued giving authority to fir Richard Bingham then governor of Conaught, and twenty-one other Commiffioners to make a compofition between the Queen and the lords and and their tenants of that country, and of Thomond, for a rent certain to be paid out of every quarter of land therein, in lieu of all manner of uncertain ceffes, cuttings, and other exactions, accuftomed to be borne to the Queen and her predeceffors for the martial government thereof; and further the Commiffion empowered them to do all things as to their difcretion fhould feem beft, as well in the faid compofition, as in the divifions of baronies into manors, and to advife all other things that fhould tend to the general good and quiet of the country, and the good fubjects of the fame.

Commiffion to make a compofition, between Q. Eliz. and the lords of Conaught and their tenants.

And accordingly, indentures bipartite were entered into on the 2d of September following, whereby it appears, 1ft. that the Lord Deputy Perrot did covenant on behalf of the Queen, that the chieftains, gentlemen, freeholders, and inhabitants, their heirs and affigns, fhould from the date of the faid indenture be exonerated for ever from all ceffes, exactions, cuttings, impofitions, purveying, catings, finding or bearing of foldiers, and from all other burdens, other than the rents, refervations and charges in the indentures fpecified, and to be enacted in parliament. In confideration whereof, the faid chieftains, gentlemen, freeholders, and inhabitants did grant to the faid Lord Deputy and his heirs, to the ufe of the Queen, her heirs

Indentures of compofition, in purfuance thereof.

and

and fucceffors, a yearly rent charge of ten fhillings
fterling out of every quarter of land within that province.
2dly. they agreed not only to anfwer for ever to all
hoftings, roads, and journies within Conaught, where
and when they fhould have notice from the government,
50 able well armed footmen, upon their own charges,
befides the rent aforefaid; and to all general hoftings
proclaimed within the realm, 20 well armed footmen,
furnifhed with carriages and victuals, at their own cofts,
during the time of the faid general hofting, if the government require it. 3dly. That the ftyles and titles of captain-
fhips and taniotfhips, and all other Irifh jurifdictions,
together with all elections and cuftomary divifions of
lands, fhould be abolifhed, and that the lands and inheritance fhould lineally defcend according to the courfe of
common law. 4thly. That the chieftains, gentlemen, and
inhabitants, fhould by letters patent have diverfe lands in
the indentures fpecified to them and their heirs, free from
the compofition, to be held by common knight fervice.
And that they fhould have all goods and chattels of
felons, and other cafualties and amerciaments.

Errors in
them.

Letter of King
James I. to
hold a com-
miffion of
grace.

But it being afterwards found, that there were various
errors and defects in the faid indentures of compofition,
and many erroneous proceedings in the execution thereof,
in order to rectify thefe errors, and to remove all doubts
concerning the aforefaid compofition, and the non-per-
formance thereof, and to remedy all the defects which
might be in the feveral titles which were derived under
the faid compofition, his Majefty King James I. by his
letter dated 21ft. July 1615, empowered furrenders to be
taken from the faid inhabitants, and gave the following
directions;

Firft,

" 1ſt, To inquire by commiſſions, what quantity of
" lands every of the ſaid inhabitants were ſeized of, and
" upon return thereof, to accept ſurrenders of ſo much
" thereof as the ſaid perſons ſhould offer to ſurrender, and
" to cauſe letters patent to be made thereof, with a
" reſervation to the King, and his ſucceſſors, of the ſaid
" compoſition royal mentioned in the indentures of Queen
" Elizabeth, and ſuch other rents and duties as were then
" anſwered to the King, to be holden by common knights
" ſervice, with a clauſe that no mention ſhould be made
" of the ſurrenders.

" 2dly, That in the ſaid grants ſhould be contained
" ſeveral pardons and grants of their ſeveral intruſions,
" fines for alienations without licenſe, meſne profits,
" reliefs, ſums for reſpite of homage, concealed wardſhips;
" or that the deputy give them ſuch other effectual diſ-
" charges as ſhould free them and their heirs from all
" future trouble in any of the King's courts; with a
" proviſo, that they ſhould firſt make ſome moderate
" compoſition for their ſaid ſeveral intruſions with ſuch
" patentees, or their aſſigns, to whom any grant had
" been made of the ſame; the fourth part of which
" compoſition was to be reſerved to the King". And
afterwards ſeveral ſurrenders were made and letters
patent granted in purſuance of the ſaid letter.

But towards the end of the reign of King James I. it
being diſcovered that neither the ſurrenders or patents
which had been ſo made and paſſed, had been enrolled in
Chancery, by which means the title of the patentees became defective, and the lands were ſuppoſed to remain
ſtill veſted in the Crown, the King propoſed to make a

New defects and omiſſions diſco-vered.

plantation

plantation there, as had been done in Ulster; though the omission was not so much the wilful default of the patentees as the neglect of the officers, to whom they had paid near £2000 for the enrolments of the patents, which were never made.

New composition between king Charles I. and the patentees.

But King James died before he could complete this scheme. And in the third year of the reign of his successor Charles I. a treaty was set on foot and concluded between him and the patentees, by which, in consideration of a sum of £120,000, agreed upon to be paid by them to his Majesty in gales, they were admitted to enrol the surrenders and patents made to them; and such as had a mind to make new surrenders, were to have the same accepted and enrolled, and new patents passed to them. And for their further security, their several estates were to be confirmed to them and their heirs by the next parliament, to be held in the kingdom.

And the patents confirmed by act of parliament.

And several acts of parliament were afterwards, though not without much difficulty and opposition through the arbitrary councils of Lord Strafford, made in this kingdom, for the purpose of confirming patents passed for those and other lands, under commissions of grace, as they were called.

How these rents are in charge in the King's books.

And the aforesaid composition rents are in charge in the King's rent rolls as follow, to wit, for every quarter part of a town land, 10s. And for every cartron, 2s. 6d. But I do not find that it was ascertained what number of acres any of the said denominations should contain.

There

There were alfo other compofition rents which depended on a compofition made by the Lords of the pale, and the inhabitants of the province of Munfter, with Sir William Fitzwilliams, who was alfo Lord Deputy here in the reign of queen Elizabeth, after Sir John Perrot; but there does not at this day appear any diftinct account of thefe rents; and it is imagined they have paffed under the denomination of quit rents.

Other compofition rents

And the amount of thefe compofition rents is about £1000 a year.

Compofition rents 100cl. a year.

Quit rent is a rent which arofe and was induced in this kingdom after the rebellion in 1641, by the acts of fettlement and explanation. And it is an acrable rent, according to the Englifh ftatute meafure, referved upon all the eftates in Ireland, which were forfeited by that rebellion, and granted by the Crown to adventurers, foldiers, and debenturers; and on lands which were then feized, and afterwards reftored to innocent papifts by decrees and certificates; or on lands given to them as reprifals; or to tranfplanters.

Quit rents, what, and when and how induced.

And the rates according to which thefe rents were referved were as follow, viz.

		l.	s.	d.
For every acre in	Leinfter - - - -	o	o	3
	Munfter - - -	o	o	2¼
	Ulfter - - -	o	o	2
	Conaught - - -	o	o	1½

And the yearly amount of thefe quit rents is about £50840.

Amount of quit rents.

G 2   Thefe

The King's rents to be paid to the collectors of the revenue.

These rents are paid to the several collectors of his Majesty's revenue, whose receipts, by 9 W. 3, 31. are good and valid in law against the Crown, and as effectual, to all intents and purposes, as an Exchequer acquittance duly passed and entered in the several offices of the Exchequer.

Collectors to give receipts for all payments, &c. and their fees.

And the collectors are thereby required, upon payment of any part thereof, to give to the person so paying the same, a full receipt or acquittance for what he shall receive, in parchment under his hand, wherein he shall mention the sum so by him received, and for what gales rent, and for what land, and on what account the same is paid to him; for which acquittance they are to receive, for any sum above five shillings, and not exceeding twenty shillings, sixpence; and for every sum above twenty shillings, and not exceeding five pounds, one shilling; and for every sum above five pounds, and not exceeding fifteen pounds, one shilling and sixpence; and for every sum above fifteen pounds, two shillings; and in no case to receive any more for one acquittance than two shillings.

Persons charged with several distinct sums in respect of several parcels of lands to have one acquittance only.

And by Sect. 4, where one person stands charged with the payment of several and distinct sums, in respect of several parcels of lands, or where the same is in charge in the name of other persons, not in possession of such lands, the several collectors are thereby required, on receipt of the said rents, or any part thereof, to give to the person in possession and paying the same, one acquittance for what he shall pay; which acquittance shall distinctly mention as well the lands and tenements, as the rent paid, and for what gale the same is paid, and by whom; for which one acquittance

quittance the collectors are to receive no greater fee than as aforefaid.

By ftat. 11 Wil. III. ch. 2. Eng. the forfeited eftates in Ireland fhall, after fale thereof, be fubject to fuch crown rents, quit rents and chiefries, as the fame were liable to on the 13th of February 1688.

Forfeited eftates in Ireland, after fale, liable to fuch quit, &c. rents as they were before 13th Feb. 1688.

But it is thereby provided that nothing therein contained fhall make void any grant of any quit rent, or other rents, made in confideration of any juft debts releafed to the crown, to the full value of fuch grant; or make void any grant for reduction, or abatement of any quit rent, where the fame abatement hath been made in confideration of the barrennefs or coarfenefs of any lands out of which the fame is iffuing, or for the better improvement thereof.

Not to extend to make void any grant of quit or other rent made in confideration of any juft debts releafed to the Crown.

By 1ft Ann. ftat. 2. ch. 21. Eng. The truftees for forfeited eftates, or any feven of them, with the confent of three of the Commiffioners of the revenue in Ireland, may apportion any quit rent, crown rent, or compofition rent, payable to the Crown, and charge the fame in parcels upon the lands liable thereunto, fo that every part fold by itfelf, having regard to its quantity and value, may be liable to a certain proportion of the faid rents; which apportionment, fet down under the hands and feals of the faid truftees, &c. and Commiffioners, &c. and enrolled in the Exchequer in Ireland, fhall be good in law, and be referved to her Majefty, &c. in the purchafe deed; and the premifes fold difcharged from the reft of the

Truftees of forfeited eftates in Ireland to apportion quit and other rents of the fame parcels &c.

the faid rents, and fhall be liable to the faid referved rent in the fame manner as it was before to the whole rent. And if the truftees and Commiffioners fhall not agree in the faid apportionments, before the 24th of May 1703, or fhall fooner difagree in them, the Chief Governor of the faid kingdom may make the like apportionment, which fhall be of the fame force as if made as aforefaid *.

Arrears from 25th March 1692 to 25th March 1695 difcharged.

By ftat. 9. Wil. III. ch. 29. and 2. Ann. ch. 4. the feveral arrears of quit rent out of lands, &c. returned to have been wafte from the 25th of March 1692, to the 25th of March 1695, are difcharged, &c.

Plus acres what, in whom vefted and to what quit rent liable.

By ftat. 2. Ann. ch. 8. the lands called plus lands, or plus acres, which are parcels of denominations of

* On the plan of this power in the truftees of apportioning rents, which expired as is above mentioned on the 24th of May 1703, the court of Exchequer I ave fince proceeded in the apportioning of rents; but always with a *falvo jure coronæ*.

But for this purpofe, a petition is to be preferred to the court, verified by affidavit, (notice being firft given to the folicitor for the King's rents.) who refer it to the Auditor general, and this order of reference is to be ferved on all the parties concerned; and when the Auditor general has apportioned the rents, his report thereon will, on motion of counfel, be referred by the court to the Attorney general, (notice being firft given to the folicitor for the King's rents as aforefaid.) who is alfo to make his report, which is generally a tranfcript of the Auditor's report; and upon his report an order is to be obtained for confirming it, unlefs caufe, which order is to be ferved on all the other proprietors of the lands; and if no caufe be fhown to the contrary, on affidavit of fuch fervice, and certificate from the fecond Remembrancer of no caufe, and counfel's motion thereon, the report of the Attorney general is confirmed abfolutely. And note, caufe may be fhown either on affidavit, notice and motion, or by exceptions to the report, which are to be fet down to be argued, as in other cafes of exceptions to his report. See the cafe of his majefty againft Nutley, 30 June 1731. Same againft Daly, 11th December 1731, againft Morgan, 31ft May 1745, and againft Rielly, Hill. 1746.

lands

lands undifpofed of, where the refiduc of fuch denominations have been granted by patent, are vefted in fuch
perfons, who on the ift day of October 1702, were in
pofleffion of fuch plus acres by themfelves or thofe deriving under them, in their right, or under colour thereof; which perfons fhall hold and enjoy fuch plus acres
to them and their heirs, liable to fuch quit rent for the
fame *pro rata*, as is payable out of the other part of
fuch denominations.

And by the faid act, reciting that there were feveral
denominations of land entirely undifpofed, which,
as alfo fome of the aforefaid plus acres, were fo coarfe
and barren that they were not worth the quit rent
they were liable unto, and therefore remained defolate,
it is made lawful for the Chief Governors and fix more
of the privy council to demife or grant the fame to fuch
proteftants, and for fuch term of years, as they fhall
think fit, at fuch a reafonable rent as may encourage
fuch perfons to plant and inhabit the fame.

But it is thereby provided that nothing therein contained fhall avoid any fettlement, leafe, charge, or other conveyance, or encumbrance made by the perfons whofe
eftates or pofleffions are thereby confirmed, or by the
perfons under whom they derive; but that the fame and
all other rights and titles, except of her Majefty, to the
faid lands, fhall be of the fame effect as if the faid act
had never been made.

Rents are ufually put in charge two ways; to wit, by
the Auditor general, *ex officio*, from the King's grant; or
by the court of Exchequer upon a *fcire facias* on behalf of
the Crown.

In

By the Auditor general, *ex officio*, from the King's grant.
In the firſt cafe, where inquifitions have been taken either by fpecial commiſſion, or by the eſcheator *ex officio*, and a particular thereof made out by the Surveyor general and Auditor general, a *fiat* is then prepared by the Attorney or Solicitor general for the patent or grant of the lands; which *fiat* is to be lodged in the rolls office; and when the grant is fealed and enrolled, it is not given directly to the party concerned, but brought, by one of the clerks in the rolls office, to the Auditor general, to be by him entered, who from thence afcertains the rent, and inferts the fame in the roll of the King's rents; and the grant is then delivered to the party.

Or on a ſcire facias for his Majeſty.
But where any old rent is to be put into charge, which hath not before been in charge in the King's rent roll, or where there is any difficulty with regard to the right, fo that it may be neceſſary that evidences and proofs be produced for the determining thereof, in fuch cafes the proper method of proceeding is to fue a *ſcire facias* for his Majeſty, directed to the ſheriff of the county where the lands lie, to give notice to the party to anfwer the charge thereon *.

The proceedings on fuch ſcire facias.
And the appearance to this *ſcire facias* is to be entered in the rule book in the fecond remembrancer's office, from whence this writ is to iſſue. And on this *ſcire facias* rules to plead are to be entered, and fuch proceedings had, as on *ſcire facias's* on recognizances to the King.

---

* So determined in the cafe of his Majeſty againſt John Daly, in this court, Hillary term, 1752. And in the cafe of his Majeſty againſt O Brien, afterwards Earl of Thomond, in Hillary term, 1758, after many proceedings had been for feveral years in each cafe, on the Attorney general's report and exceptions thereto.

It

It has often happened that the same lands have been
granted by patent to different persons; in which case, if
it were upon the acts of settlement and explanation, the
lands were adjudged to the patentee who had the prior
certificate; but yet as these lands were often mixed with
other denominations in the same patent, the patentee,
who did not enjoy the lands so doubly granted, was never-
theless liable, as to his other lands, to the rent reserved
upon the lands which he did not possess.

Disclaimer, in what cases.

In all cases of this sort the grantee of the lands is to file
a plea of disclaimer in the second remembrancer's office,
wherein he is to set forth his title to the lands in the grant
which he doth possess, and then to disclaim the lands
which he doth not possess; and to this plea the Attorney general is to file a confession; and thereupon the
court will give judgment for exonerating the defendant,
and the lands of which he is seized, from the rents so
charged upon the lands he hath disclaimed; and this
judgment is accordingly made up and enrolled among the
records of the said office.

and how.

As to over charges and double charges * which have
happened in several cases, as where more acres have been

Over charges and double charges in several cases.

---

* In the year 1758, it appearing to the commissioners of the revenue that several
lands had, for many years, been returned by the several collectors of the kingdom,
in their lists of arrears, as double charges, and that several persons had, from time
to time, been grievously vexed by distresses and otherwise for the same, on the 11th
of July in the same year an order of the board was made, that the solicitor for the
King's rents should cause all the said double charges to be discharged and struck out
of the rent rolls at the expense of the Crown, and without any expense to the
subject; which was accordingly done on constats thereof from the Auditor general, the consent of the Attorney general, and on motion thereon, by the solicitor
for the King's rents without any other or further proceeding.

diftributed and granted, than the lands forfeited, by the furvey and diftribution book, appear to contain; or where there have been two grants at different times to the fame perfon, or thofe deriving under him, of the fame lands, and the rents referved on each are both continued in charge, whereas the laft rent only ought to be charged; or where the fame lands are charged in two diftricts, or charged twice in the rent roll; or where part of the lands lie in one barony and part in another, but being in different diftricts are charged in the rent rolls in both; or where a perfon apprehends that the Crown has not a right to any rent with which his lands are charged, or that the lands are not liable thereto; in fuch and the like cafes the party fo charged may, as well before as after any diftrefs or other proceeding is taken for the recovery thereof,

The proceedings to difcharge them by petition. move the court by counfel on a petition fetting forth the facts, (which are ufually verified by affidavit) that it may be referred to the Auditor general to report thereon; and the court will make an order for that purpofe, with a refpite of proceedings in the mean time : but the ufual notice of this motion is to be given to the folicitor for the King's rents. Then an attefted copy of the order is to be brought to the Auditor general, who thereupon will, if required, iffue fummonfes for attendances on him, and proceed to make his report.

And the Auditor's report referred to the Attorney general. And when the Auditor general has made his report, the party, in whofe favour it is, may move the court by his counfel that the Auditor's report may be referred to the Attorney general, who makes his report thereon, which is ufually *totidem verbis* with the report of the Auditor general [*], (for which his fee is two guineas); then a motion

---

* Of late the Attorney general examines into the matters contained in the report of the Auditor general, in prefence of the folicitor for the Crown and the attorney for the petitioning party, and varies the Auditor's report as he fees fit.

may

may be made by counfel, without notice, to confirm the Attorney general's report; and the court will make an order for that purpofe, unlefs caufe be fhown to the contrary in four days after fervice of the order. And if no caufe be fhown in thefe four days, the court, on certificate thereof and affidavit of the fervice of the order, will on counfel's motion make the order abfolute, and confirm the report. And if the report be in favour of the fubject, the court will alfo order the rent and arrears to be ftruck out of charge; and if a diftrefs be taken, or fecurities or money lodged in the hands of the collector or other perfon, they will order them to be reftored to the party.

And either party may, if he thinks himfelf aggrieved by the report made by the Attorney general, take exceptions thereto; and thefe exceptions are to be fet down as a caufe, and argued in court by the counfel for the Crown and party; but no exceptions are to be taken to the report of the Auditor general, as his report muft be referred to the Attorney general *.

Exceptions may be taken to the Attorney general's report, but not to that of the Auditor general.

But if it be a matter of great nicety and difficulty, the court will not determine it upon thefe exceptions; but will leave the party to be relieved by *traverfe*, *fcire facias*, &c. according to the circumftances of the cafe. See the cafe of his Majefty againft the tithes of Jerpoint in this court, Trinity Term 1749, where in fuch cafe a *traverfe* was granted to the fubject, although the King was in poffeffion by virtue of an injunction on a cuftodiam, on giving fecurity by recognizance to be anfwerable for the poffeffion and mefne rates, if adjudged againft the

---

* So determined in the cafe of his Majefty againft the right honourable Percy Wyndham O'Brien afterwards earl of Thomond, in Hill. Term 1758.

traverser, &c. and afterwards an *amoveas manus* was granted.

Petitions to the Exchequer for discharge of arrears of quit, &c. rents, where none have been paid for 20 years before, &c.

By stat. 3 Geo. III. ch. 22. It is made lawful for every person, bodies politick and corporate, at any time before the 25th day of March 1770, to prefer their petitions to his Majesty's court of Exchequer, thereby setting forth, that all or some of the lands, rectory, abbey, priory, or monastery lands, tithes, fairs, tenements and hereditaments whereof they are seized, are subject or liable to some certain quit rent, crown rent, composition rent, or other chief rent, payable to his Majesty, which hath not been paid by them, or those under whom they respectively derive, for 20 years next immediately preceding the 29th day of September 1764, particularly describing, in such petition, the lands liable to the payment thereof, as well by their present as former names and denominations, and thereby submitting to pay all such annual quit rent, crown rent, composition rent, or other chief rent, as shall become due after the said 29th day of September, and praying to have such lands discharged from all arrears of such rent incurred, due before the said 29th day of September 1764; on which petition an order shall be made by the court, that the Auditor general shall search into the respective rent rolls, books and records in his office, and shall certify to the court, by a certain day to be appointed by the said court, whether any such rents, as are mentioned in such petitions, have been accounted for to his Majesty or his predecessors, within the space of twenty years next preceding the said 29th of September, 1764.

And the Auditor general is thereby required to certify to the court, whether it appears to him that such rents have

have been paid or accounted for to his Majefty or his predeceffors, within the fpace of 20 years next before the
29th day of September 1764; for which fearch two
fhillings and fix pence and no more, and for which certificate fix fhillings and eight pence and no more, and for
entry of a difcharge of fuch arrears out of the rent rolls
three fhillings and four pence and no more, fhall be paid
to the faid Auditor general.

And if upon return of fuch certificate, and upon
examining into the truth of the allegations of fuch petition, by the court, in a fummary way, it fhall appear to
the court that the allegations contained in fuch petition
are true, or if it fhall appear that no quit rent, crown
rent, compofition rent, or other chief rent payable to his
Majefty, has been paid for or out of fuch lands, or has
been accounted for to the collectors of his Majefty's
revenue, for the diftrict wherein fuch lands lie, within the
term of 20 years next before the 29th day of September,
1764, in fuch cafe the court is by faid act required to
make an order on fuch petition, that the lands therein
mentioned and the perfons who from time to time refpectively held and enjoyed the fame, fhall be abfolutely freed
and difcharged of and from all fuch rents and arrears due
or in arrear at any time before the faid 29th day of September, 1764; and fhall order the faid Auditor general to
give in charge fuch growing rents as the faid lands fhall
appear to be charged or chargeable with, which fhall
become due from and after the faid 29th day of September,
1764, to the collectors of the diftricts where fuch lands
lie, to the intent that the fame may be duly collected for
the future; and fuch order fhall be an effectual difcharge
againft his Majefty, as to fuch publick arrears.

The

Old method
of collecting
the king's
rents.

The old method of collecting thefe rents antecedent to the year 1693 * (at which time the eftablifhment of the kingdom was fettled) was thus ;

The patents on which thefe rents are referved were originally, after they were enrolled, (as they are at this day) entered with the Auditor general of the Exchequer, who out of them made an abftract of the rent referved, for what land, in what country, from whom, and of the date of the patent ; and thefe abftracts were reduced under the heads of the feveral counties where the principal denominations of lands, out of which the rents were referved, lay.

By procefs of
the pipe to
the fheriff.

From this office a rent roll was twice a year tranfmitted, fometimes to the fecond remembrancer's office, fometimes to the clerk of the pipe; which laft office, twice a year, viz. at Michaelmas and Hillary terms, made out procefs to the refpective fheriffs of the feveral counties in this kingdom for the collection of them.

Who ac-
counted for
them.

And the refpective fheriffs for each county, once a year when they paffed their accounts, accounted with the court of Exchequer for thefe rents ; and the whole rent roll for each county being in open court read over to the fheriff, he upon his oath gave his anfwer to each particular ; if received, he charged himfelf with it, which in the Exchequer language is called *Tot*, i. e. *totum in manibus* ; if not received, he gave his reafons upon oath why he

* The total amount of thefe rents at that time was £65052 16 3, out of which £1664 10 were to be deducted for feveral of the faid rents granted to particular perfons by K. Ja. II and K. Wil. and Q. Mary ; but they were afterwards made unalienable by the Englifh act of 11 Wil. III. as is before mentioned.

could

could not collect it; which, if juft, were allowed by the court; if not, the court charged the fheriff with it, and let him take a writ of affiftance to collect it for himfelf.

The arrears were fent out again by writ or by procefs to the fucceeding fheriff together with the rents of his time.

The fheriff took a *debet* of the balances due on his account from the clerk of the pipe, and thereupon paid the money into the treafury.

And took a debet from the clerk of the pipe, &c.

This method of collecting was afterwards thought inconvenient for the following reafons, viz.

This method inconvenient.

1ft, The rents came in but once a year, the fheriffs not being to account oftener; and it was not convenient that fo large a fum as £30,000 or upwards, which was the firft half years rent after the eftablifhment was fettled, fhould lie in the hands of the feveral fheriffs of the kingdom until the year was out, when the King's occafions required it before.

The reafons why.

2dly, They were never well brought in by the fheriffs; for they not being able to collect them themfelves but doing it by their bailiffs, feveral fums which were received were, either by neglect or out of defign, not returned by the bailiffs as received; fo that, the fheriff not charging himfelf with them on his account, the fubject, on the renewal of the procefs, was forced to apply to the Exchequer for redrefs to his great expenfe.

3dly, Great arrears were found to be returned by the fheriffs, they either out of favour to particular perfons,

or

or regard to perfons of quality, neglecting to levy the rents from them.

4thly, Great fums licing in the hands of the fheriffs or fub-fheriffs until their accounts were paffed, they were thereby tempted to fpend or mifapply the money, which proved often the ruin of themfelves and their fecurities.

Befides the charge of iffuing it in this method was inconvenient to the fubject; for it generally happened that lands in feveral denominations, in diftant places, fometimes in feveral counties, were paffed in the fame patent, and a certain yearly rent referved to be iffuing out of the whole; in which cafe, each parcel of land being by law liable to the whole rent, the Auditor general, in his rent roll or charge for the fheriffs collection, had given directions for demanding and levying the whole rent only on the principal denominations of land in the patent, no notice being therein taken of the other denominations which were in the hands of other perfons and often in different counties; by which means the tenant of the principal denomination was forced to be at great trouble and expenfe to get in the proportions of the tenants of the other lands.

This being the general cafe was very mifchievous, and not to be avoided but by apportioning the rent on each denomination of land in charge in the grant, and then iffuing directions accordingly for the collecting from the particular tenants of each denomination their feveral portions, which was without prejudice to the King's remedy upon the whole, in cafe any particular parcel of land fhould become unable to difcharge its portion of rent.

Thefe

These are the reasons commonly given for changing the method of collecting by sheriffs; but these mischiefs might in a great measure have been remedied without altering the course. And there is one more which was thought to be of greater force, and the chief motive of changing the old officers for collecting and managing the revenue through most of its branches. In the year 1669 the whole revenue of this kingdom was set to farm to John Forth, of the city of London, alderman, and ten others, for seven years from the Christmas before, at several distinct yearly sums amounting in the whole to £219,500. As soon then as the farmers had got the receipt and management of the revenue into their hands, it was concluded, and with good reason, that officers who depended on them and the commissioners of the revenue for their offices, and who were liable to be removed on the least apprehension of neglect or other default in them, would be more strict and circumspect than such as had no dependance on them, and whose office was rather a burden to them without profit, as the sheriffs collection was; and who had legal estates by patent in their offices, and whom consequently the farmers and commissioners of the revenue could not control.

The principal reason for changing the old method of collection.

And therefore the whole kingdom was divided into several districts, according as the land lay in compass, for the ease and conveniency of the collection; the division by counties being unequal; some of them being of too great an extent for the collection, and some too small; and a particular collector was appointed for each district.

New method by dividing the kingdom into districts and appointing collectors.

And as the King received a certain rent from the farmers, and was not concerned how much the revenue

A clerk of the quit rents and Accomptant general.

VOL. I.   I   produced,

produced, the farmers were suffered to manage it their own way; who thereupon slighted the course of the Exchequer, and appointed their own officers; viz. a clerk of the quit rents, who made out the charge to the collectors, and an Accountant general, before whom these collectors were to account; both which served instead of the Auditor general. And these officers made out their charge, and settled their books by the records in the Auditor's office, of which the farmers, by their patents, were to have the inspection and use during their continuance.

Subsequent alterations in the rent roll not taken notice of by the Auditor.

But several alterations having been made in the quit rents, some upon commission for reducing them in several places where they were too high, and others upon orders from the Exchequer, the alterations were entered with the farmers clerk of the quit rents, but not with the Auditor, nor any notice taken of them in his book. Besides, the change in issuing of the charge from counties to districts, and the apportionment of the rents, made a vast alteration in the rent roll; so that by these means, at the end of the farms, the Auditor was not able to make out any charge for the collection of the quit rents, nor was there any certain rent roll or record for that part of the revenue, but it lay wholly in the breast of the clerk of the quit rents to the farmers, and his private books, on which there was no cheque.

After the revolution a new rent roll made.

It continued thus after the end of the farms until the revolution; but since that period the court of Exchequer, taking notice of the condition which that part of the revenue was in, thought it necessary that there should be a certain rent roll or record of the quit rents, to remain as a charge to the collectors, and a cheque to the persons concerned in the receiving the same; and did, with

with the encouragement and affiſtance of the then government, out of the quit rent books and Auditors
books, and by comparing them with the original patents,
fix a certain and methodical rent roll of the quit rents,
according to the preſent courſe of collection by diſtricts,
in the manner following viz.,

Under the head of each diſtrict are placed the particu- In what
lar counties; and under them the baronies, tenants names, manner.
denominations of land, the number of acres of the whole,
and the apportioned rent of each denomination, with
the reduced rents, if there be any * reducements, each
in diſtinct columns.

The rent roll for each diſtrict, according to its form, Which iſſues
iſſues out, once a year, to each collector, and is his once a year
charge; and he collects theſe rents, and gives an ac- lector.
count to the commiſſioners of this and the other branches
of the revenue, which he pays into the treaſury, from
time to time, as there is occaſion. And between Lady

* Theſe reducements were, by virtue of a commiſſion under the great ſeal of
Ireland, directed to his Excellency, Arthur Earl of Eſſex, then Lord Lieutenant of Ireland, and others, bearing date the 25th June 1676, grounded on his
Majeſty's letter under his royal ſignet, dated at Whitehall the 3d of December
1675, whereby they were empowered to reduce and abate quit rents and arrears
thereof, due out of coarſe and barren lands in the Kingdom of Ireland,
where the quit rent was equal to or near the yearly rent of the land. And
theſe reducements were engroſſed on rolls of parchment, and are in the Chief
remembrancer's office, and are readily reſorted to by means of the name of the
perſon in whoſe favour the reduction was made; and other reducements were
made ſeveral years before, by the Lords Juſtices of this Kingdom, by virtue
of the King's letter, upon petitions of officers, ſoldiers, and others, to whom
ſuch coarſe and barren lands had been ſet out. And the orders for theſe reducements are in the offices of the Auditor general and Surveyor general; and
theſe matters were referred to theſe officers for their report, before the orders were conceived.

I 2   day

day in March, and Lammas in every year, the feveral collectors of the kingdom pafs their accounts for the whole year, according to the time appointed by the commiffi ners for that purpofe, in the manner herein after mentioned.

C H A P. V.

OF THE CUSTOMS, AND IMPORT EXCISE, AND ADDITIONAL DUTIES ON GOODS IMPORTED, AND EXPORTED, AND THE MANNER OF COLLECTING AND ACCOUNTING FOR THEM.

Cuftoms, what.

THE cuftoms are the duties of *poundage* and *tunnage* on goods imported, and of *poundage* on goods exported.

Poundage old.
Davies 31.

Poundage is an ancient duty, payable to the crown on all merchandize and wares imported into or exported from this realm, to be fold; except wines and oils, which pay cuftom by way of tunnage. And this duty, which has been granted to the crown, by various acts of parliament, in England, from the reign of Edw. III. and moftly after the rate of twelve-pence in the pound, according to the feveral and refpective values and rates of the merchandize, is faid by Sir John Davies to have been firft granted to King Hen. VII. in this kingdom in the 10th year of his reign for 5 years; and at the end of that term to him and his heirs for ever, after the rate aforefaid. And this is called the old poundage.

---

* The word cuftom, which is denominated in the ancient barbarous latin *cuftuma*, and not *confuetudo* (ufage) feems to be derived from the french word *couftum*, or *coutum*, which fignifies toll or tribute, and owes its own etymology to the word *couft*, which fignifies price, charge, or *coft*. 1. Blackft. Com. c. 8.

And

And by the 14th and 15th Car. II. c. 9. the said duty of and new. 12 pence in the pound on all goods imported and exported, (except wines and certain oils) is granted to the King and his heirs, to be paid according to the several and particular rates and values of such merchandize, as they are refpectively rated in the book of rates annexed to the faid ftatute. And in cafe of importation or exportation of any goods not mentioned therein, the poundage is directed to be levied according to the true value, to be affirmed upon the oath of the merchant, in the prefence of the cuftomer, collector, comptroller and furveyor, or any two of them.

But out of this duty of poundage on goods imported Allowance out of it. there is to be, by the 8th rule of the faid act, an allowance of £5 per £100.

Tunnage is a duty payable by the faid ftatute on wines Tunnage. and certain oils imported into this kingdom, viz.

		l.	s.	d.
For every tun of French wine imported				
	by fubjects,	3	10	0
	by ftrangers,	4	13	4
For every pipe or butt of Levant, Spanifh, or				
Portugal wine,	by fubjects,	2	10	0
	by ftrangers,	3	6	8
For every awme of Rhenifh,	by fubjects,	0	15	0
	by ftrangers,	1	0	0
For every tun of rape and linfeed oil,				
	by fubjects,	0	15	0
	by ftrangers,	1	0	0
For every tun of Spanifh &c. oil,	by fubjects,	2	12	0
	by ftrangers,	3	5	0

For

For every tun of fallet oil,   by fubjects,   3   3   0
                                    by ftrangers,   3 18   9
For every tun of oil of Greenland, by fubjects,   0   8   0
                                    by ftrangers,   0 10   0
For every tun of oil of Newfoundland,
                                    by fubjects,   0   6   0
                                    by ftrangers,   0   7   6

Allowance out of it for leakage. But out of this duty there is to be, by the 7th rule of the faid act, an allowance of £10 per £100, for leakage on all wines imported; provided fuch wines have not been filled up on board the veffel.

Goods wrecked not liable to cuftom.
Vaugh. 159.
L. Raym. 388.
501.
It has been determined that goods fhipped in foreign parts as merchandize, and wrecked on the coaft, are not liable to the duties impofed by this act; as they could not be deemed imported within the meaning of it.

But goods faved, not being wrecked, liable. But by the 6 Geo. I. c. 8. all goods which fhall be faved out of any veffel that fhall happen to be forced on fhore or ftranded on the waftes of this kingdom, not being wrecked goods, *jetfam*, *flotfam*, or *lagan*, fhall after all charges of falvage &c. be fubject to the payment of cuftom as if imported.

Impoft excife, what The *impoft excife* * or *new impoft* is a duty of poundage granted by 14 and 15 Car. II. c. 8. to the King and his heirs on all commodities, merchandizes and manufactures imported, (jewels, bullion, corn, arms and ammunition

---

* So called from the Dutch word *accife*, which fignifies an affeffment upon a commodity; or from the word *excifum*, a part of the profit cut off from the whole. Gilb. Treat. of Exch. 252.

excepted)

excepted) according to the rates they are valued at in the book of rates annexed to the said flatute, viz. for all forts of drugs 2 fhillings in the pound, for all forts of raw hemp, undrefs'd flax, tow, rofin, pitch, wax, cable, cable yarn or cordage, 6 pence in the pound.; for all wines, tobacco, falt, and other goods fpecified and valued in the faid book of rates, one fhilling in the pound; and for all other goods not fpecified or rated in the faid book of rates, one fhilling in the pound according to the book of rates for cuftoms; and if omitted there, then as they fhall be rated and valued by the fub-commiffioner collector and fearcher for excife in the place where imported, or according to the higheft market price.

All which duties are to be paid by the firft buyer before his receiving them from the merchant importer; unlefs the merchant be a fhop-keeper, retailer, or one employing

To whom to be paid.

* It is this impofition of the duty upon the buyer that conftitutes the effential difference between cuflom and excife, properly fpeaking; the former being a tax immediately paid by the merchant, altho' ultimately by the confumer; the latter, an inland impofition paid either upon the confumption of the commodity, or upon the retail fale, which is the laft ftage before the confumption And the excife is doubtlefs the moft œconomical way of taxing the fubject, and renders the commodity cheaper to the confumer; for this obvious reafon, that the earlier any tax is laid on a commodity the heavier it falls upon the confumer in the end; becaufe every trader, thro' whofe hand it paffes, muft have profit, not only upon the commodity, but alfo upon the tax itfelf. But this good effect of the excife is not produced by this claufe of the act which brings the duty one ftage nearer the confumptioner; it being now generally paid in the firft ftage, or by the merchant importer, on the terms herein after mentioned. So that the diftinction in this kingdom between the cuftom inwards and import excife anfwers no other purpofe, than to make the collection of thofe cuftoms much more intricate and complicated than it would be if they were both granted by one undiftinguifhing law, and levied in one manner, and by the fame rate, under the general term of cuftom And this diftinction likewife furnifhed a ftrong objection againft the late meafure of dividing the boards of cuftoms and excife; from the delay and additional expenfe which muft arife to the merchants by the neceffity of returning feparate accounts to the two boards.

them

them for his own confumption; in which cafe the duties are to be paid by the importer, before he be permitted to carry the goods away from the cuftom houfe or place of landing.

And it is obfervable that wine pays cuftom by the mea-fure, viz. by tunnage; but excife by value, viz. by poundage.

A bond to be given for it by the importer, and an ac-count kept.

But the duty of excife not being payable until the goods are fold, a bond was by the act directed to be given by the merchant importer, conditioned not to deliver any of the goods to any of the buyers thereof, or to any fhop-keeper or retailer whatfoever, till fuch time as the excife fhould be duly paid by fuch buyer &c. and an import account was kept according to the act.

But now an allowance made for prompt pay-ment.

But this being found very inconvenient, on the 21ft October 1679 an agreement was entered into between the then commiffioners of the revenue (by virtue of a power in their patent fo to do) and feveral merchants, by which, for prompt payment of the duties of excife, allowances of £10 per £100 in the excife and additional duties on wines and tobaccoes, and of £6 per £100 on all * other goods im-ported, were given to the importer. And this agreement, though at firft but for one year and for the benefit of the merchants only who figned it, being found to be of equal benefit to the crown and fubject, became general by ufage and is continued to this day.

To wholefale merchants only.

But none but wholefale merchants are entitled to thefe deductions or allowances. Retailers or confumptioners

---

* But coals, flates, coaches, and chariots have always been deemed retailed goods, and excluded from the benefit of the £6 per cent deduction.

are

are to pay down their excife, as the act directs, without any deduction.

And the merchants who are entitled to thofe allowances are generally well known. But where the merchant is not known, (if it be in the city of Dublin) he is to produce a certificate under the hands of feveral known merchants, which is firft referred to the collector, who examines into the truth of it; and if on inquiry he finds the perfon qualified, he makes his report accordingly; and the commiffioners make an order for allowing fuch perfon the benefit of an wholefale merchant. *How obtained in Dublin.*

But if this allowance be demanded in any of the diftricts in the country, a certificate is to be produced to the commiffioners, figned by fome of the principal merchants of the city or town in which the perfon claiming it refides, and alfo by the collector and furveyor of the diftricts; which certificate the commiffioners refer to the examinator of the import excife; and on his report an order for the allowance is made out. *How in the country.*

But befides the aforefaid perpetual duties of cuftom and excife, there have been, from time to time, granted by parliament, and are now payable, various additional and temporary duties on feveral goods and merchandizes imported and exported; the principal of which are thofe on tobacco, fpirits, and wine imported. And thefe duties are collected and levied according to the act of excife, viz. may be bonded by wholefale merchants, or paid down in ready money with the fame difcount allowance. *Additional duties on goods imported and exported.*

The collection of the ancient cuftoms, antecedent to the aforefaid acts of cuftoms and excife, was in this manner. *Ancient method of collecting the cuftoms.*

K There

There were then in every port a cuftomer, a comptroller, and a fearcher; and thefe three officers took care of all goods imported or exported. They are mentioned in Cotton's records, 17 Edw. 3. fol. 38, &c. *.

Cuftomer, his duty.

The cuftomer, who in Dublin port is called cuftomer and collector, was the moft ancient and at firft the fole officer who was the collector of cuftoms, and accountable for them to the King; 9 Hen. 6. 12 b. and 1 Hen. 7. 4 b. He was to make an entry of all goods and merchandizes imported or exported; to rate or tax the original bills of entry; to receive the cuftoms; and to fign all warrants for the charging and difcharging of goods. He was to charge himfelf upon oath, and to difcharge himfelf by tallies of payment; and upon his appointment to his office he was to give fecurity before the Chief Baron of the Exchequer, for his true accounting and anfwering his balance.

Comptroller, his duty.

Then becaufe the cuftomer was accountable to the King, but could not be charged but by his own book of cockets or his oath, a comptroller was appointed as a check upon him; and his office was to rate another bill of entry of all goods imported and exported; to take an account in his

---

* Thefe feem to be the ancient port officers, and hold their patents from the Crown; though by feveral old flatutes they fhould only be made during pleafure. And notwithftanding the alteration which has been fo made in the conftitution, conduct, and management of the revenue of Ireland, yet thefe officers are continued in feveral parts of the kingdom, and have falaries on the civil eftablifhment; to wit, there are in Dublin a cuftomer, comptroller, and fearcher; in Limerick, the like; in Waterford and Ro s, the like; in Kinfale, the like; in Youghal and Dungannon, the like; in Drogheda, Dundalk, and Carlingford, the like; in Cork, a cuftomer and a fearcher; in Galway, the like; in Carrickfergus, the like; in Strangford, the like; in Wexford, a cuftomer and a comptroller; in Killybeggs, the like; in Dinglecouch, the like.

book

book of the quantity and quality of all fuch goods, in the nature of the cuftomer's entry; and to counter-fign all warrants for the charging and difcharging of goods.

Afterwards, in aid of the cuftomer, to find concealments and fubftraction of cuftoms and fubfidies, and to feize all merchandizes forfeited, the fearcher was appointed; who in his grant of office was alfo called packer and gauger. He anciently received the warrants and cockets from the cuftomer and comptroller to unlade or lade the goods. When the merchant had paid his cuftoms to the cuftomer, he had a warrant from the cuftomer and comptroller to land the goods; but if they were landed before the cuftoms were paid or compounded for, the goods were forfeited, and the fearcher was to make a feizure of them. If the goods were to be fhipped outwards, the merchant went to the cuftomer and comptroller and entered the goods, and paid the cuftoms or agreed for the cuftoms outwards, and when fuch payment or agreement was made, they received from fuch cuftomer or comptroller a licenfe to export fuch goods, which was called a cocket. He alfo viewed all the goods and examined the feveral fpecies of them, to fee if they agreed with the warrant of difcharge, which he entered in his book.

Searcher, his duty.

Laftly, to difcover and prevent frauds in all thefe an officer called the fupervifor was eftablifhed, who is now called Surveyor general of the cuftoms of the whole kingdom. And he has deputies under him in each port, whofe bufinefs it is to furvey and overfee, from time to time, all and every the cuftomers, comptrollers, fearchers, and other officers to be employed in the ports, creeks, and havens,

Surveyor. Supervifor, his duty.

in

in and about the collecting, comptrolling, and furveying the cuftoms, fubfidies, poundage, and impofitions, and to fee that they and every of them did, from time to time, well and truly perform and difcharge their feveral duties in their faid feveral offices, in due collections and payments of the faid cuftoms, &c. and in the keeping true books and records of the fame; and in returning the faid books every half year duly into the Exchequer, in Michaelmas and Eafter term, and to the Auditors for the revenue; and to fee that the faid books and accounts of the profits thereby arifing and growing due unto his Majefty were duly audited and certified in the faid court of Exchequer; and to caufe the faid officers and every of them to do and perform all and every thing whatfoever appertaining to their faid feveral and refpective offices.

Ancient manner of accounting for the cuftoms, by iffuing port books yearly to the cuftomers, &c.

And the manner of accounting for and paying in the ancient cuftoms was thus. There iffued yearly from the chief remembrancer's office parchment books, fealed with the feal of the court of Exchequer, for each cuftomer, and for the feveral comptrollers of the feveral ports of the kingdom; on each of which books was endorfed the port for which it was; that it contained a certain number of leaves, and was for an entry to be made therein of all goods imported and exported, and the duties payable thereon, from fuch a day to fuch a day.

Who were to make returns upon oath of the cuftoms received, and to pay them into the treafury.

Thefe books were called port books; and from them the faid officers or their deputies were twice in every year, to wit, in Michaelmas and Eafter term, to give and deliver true and exact accounts to the faid court of Exchequer upon oath, of all and fingular the cuftoms, fubfidies, impofitions, and fums of money, collected and received for his majefty's ufe; and were alfo at the end of every

year

year to return the faid port books ; and the refpective cuftomers or collectors were to pay all the money they had fo
collected into his Majefty's treafury.

Thus it continued until the King's revenue was fet out   *Other officers*
to the farmers of the revenue, who (as has been faid before)   *appointed by the farmers of*
appointed officers of their own to receive the duties, inftead   *the revenue.*
of the patent officers. But yet the patent officers were neceffary at times in their employments; for there were feveral other acts to be done by them which they only can
do; efpecially where the feal of the office is required for
the more authenticating any act; as in cafes of cockets
and certificates, &c. for the commiffioners of excife have
no feal, nor of courfe the collectors appointed by them.

The act of excife and new impoft directs that there   *And by the*
fhall be commiffioners for that duty not exceeding five in   *commiffioners*
number, and a furveyor *, to be appointed by the Lord   *of excife and*
Lieutenant or other Chief Governors of Ireland, who are   *revenue.*
to be managers and governors of the office of excife created
by the faid act; with a power, with fuch approbation of
the Lord Lieutenant, &c. to appoint collectors † in the
                                                        feveral

* I do not find that any fuch officer as furveyor of the excife at large has been
ever appointed under the act. There is ftill indeed a furveyor general of the cuftoms, fubfidies, poundage, and impofitions; but the excife cannot be conftrued to
come under any of thofe denominations, it being an entire diftinct duty.

† The firft ftatute which mentions the collector is the 3 Hen. 6. c. 3. Eng. But
Lord Chief Baron Gilbert, in his treatife of the Exchequer, is of opinion that by the
collector mentioned there is not intended the collector of the cuftoms, (for that
that officer came in much later) but the collector of the fubfidy of tenths and fif-
teenths. But this feems not agreeable to the very words of the act. And the act
of poundage and tunnage, 14 and 15 Car. 2. c. 9. feems to confider the collector of
the port or collector of the cuftoms as an officer before that time exifting ; though
whether he were a diftinct officer or one and the fame with the cuftomer does
                                                                            not

feveral ports of the kingdom to collect the revenue of
excife. And by the patent of the commiffioners and governors of the revenue who are feven in number, of
whom five are efpecially to manage the excife, they are
empowered to appoint, and do appoint (amongft feveral
other officers) receivers and collectors to receive
and collect during his Majefty's will and pleafure
(amongft others,) the faid revenues arifing by the faid
act of poundage and tunnage. And thofe revenues are
accordingly collected and accounted for, and paid in by,
the collectors of the feveral diftricts together with the
other revenues received by them *.

The

not plainly appear. But it is faid that, in general, the faid collectors appointed by
the commiffioners of excife have been deputed by the cuftomer and other patentee officers to collect and receive the cuftoms; and the accounts, in which the
cuftoms and excife are blended, are always certified by the two patentee officers,
the cuftomer and comptroller; for by the act of poundage and tunnage the
commiffioners of the cuftoms have not any power to collect the cuftoms, nor
to appoint any officer for the purpofe; fuch a power would have been to the
prejudice of the cuftomer and comptroller, who had patents antecedent to the
faid act.

* Notwithftanding the cuftoms are thus received and accounted for, yet the port
books were for many years fent forth to the cuftomers and comptrollers from the
Chief remembrancer's office, who is entitled to a fee of 10 s. for each book;
but no return having been made by the faid officers for feveral years, as by their
patents, it was faid, they were bound to do, and they and the commiffioners of
the revenue refufing and declining any longer to pay for the faid books, a complaint was made thereof to the court of Exchequer, by one of the Chief remembrancer's fecondaries, on the 18th of November 1757, whereupon an order
was conceived that they fhould fhow caufe why they fhould not return the faid
books.

Upon fervice of which order feveral affidavits were made by the deputies to
feveral of the faid patentee officers, and particularly by the deputies to the
comptrollers, &c. of Londonderry and Galway, whereby it appeared that no
fuch port books had been delivered to them for many years; and upon examining of feveral of the patents it appeared, that there was no fuch claufe therein for returning the faid books as had been fuggefted.

But

The duty of the collector of Dublin port is the fame with that of the patentee cuftomer and collector, as to making entries and rating all goods imported and exported.

And after the goods have been fo entered and rated by the patentee cuftomer and collector, what they have fo done is to be examined by the examinator of the cuftoms; and if they be under rated, it is to be made good by the merchant or officer who rated the entry; if the miftake be to the prejudice of the merchant, he is to have an allowance or draw back in his next entry.

And there are alfo in every diftrict of the kingdom furveyors of feveral kinds, to wit of the port, and excife, tide furveyors, &c. who are to direct, inftruct, and infpect the feveral officers inferior to them, in the diftrict, port, or place allotted to them, or which they have the conduct and management of.

But however, the faid order of the 17th November 1757 was notwithftanding made abfolute by an order of the 11th of June 1758; and it was ordered that the feveral cuftomers and comptrollers of the feveral ports of the kingdom fhould return yearly into the Chief remembrancer's office the feveral port books they fhould be ferved with by the purfuivant of the faid court of Exchequer, or be attached without further motion; and that the purfuivant fhould deliver yearly to the feveral cuftomers and collectors the feveral port books which fhould be delivered to him by the ufher of the faid court, and that he fhould return the names of fuch cuftomers, &c. as fhould refufe to accept or receive the faid books, by the firft of the then next Michaelmas term.

Accordingly they were delivered by the ufher to the purfuivant, and the purfuivant fent them out; but he being entitled to a fee of 3s. 4d. on the delivery of each of the faid books, and the cuftomers and comptrollers refufing to pay the fame, he would not give them the books; and having made an affidavit of thefe facts, and thereupon the faid matter being on the 24th of February 1759, again brought before the court, it was ordered to ftand over until the term following; but nothing further appears to have been done therein.

And

Two princi-
pal ones in
Dublin and
their duty.

And on the cuftom houfe quay in Dublin there are two principal ones, who are to infpect the other officers employed, to inftruct them in their duty, and to fee that they do it; to take an account of all wine and tobacco difcharged; and to fee the difcharge of all goods inward and outward; and if there be any excifeable goods, not in the book of rates, to value them in the excife, after the cuftomer, comptroller, &c. have rated the cuftom thereof. And when this is done, and the proper allowances made, the merchant is to draw three entries thereof as they are called, one for the collector, who is to receive both the cuftom and the excife; one for the cuftomer and collector; and one alfo for the examinator of the cuftoms, who, if he finds any error or miftake therein, immediately puts a ftop to any further proceedings of the merchant until the fame is rectified. The furveyors on the other quays are alfo to inftruct and direct the other officers on the quays allotted to them, and to examine all boats going up and down the river, &c.

Store keeper,
furveyor and
comptroller
of the ftores
their duty.

There are alfo a ftore keeper, furveyor and comptroller in the ftores, who are to examine all fine and other goods brought into the ftores before they are difcharged, and to take an account of thofe that are not difcharged.

Tide waiters,
land waiters,
their duty.

The tide waiter and land waiter, it is faid, were formerly but fervants to the furveyor; but they are now commiffioned officers; and when the tide waiters are on fhore they are to attend at the tide furveyor's office, to be ready at all times to go with him on board fuch fhips as he fhall think convenient to place them in; and when he fo goes on board he is to rummage the fhip, and, if he finds any fine goods he is to fecure them and bring

bring them to the ftores, or to give them in particular charge to officers on board, and to fend an account thereof to the collector and land waiter; and the tide waiters fo placed on board are to take an account of all the goods, as they are difcharging, and to take fpecial care that none be concealed or fecretly conveyed away; and that they are delivered to the land waiters, who are alfo to be careful to attend the difcharging thereof on the quay, and to fee that the goods and the notes thereof agree. And if there be any fine goods or fmall parcels in any boat or lighter, the land waiters are to take care that the fame be immediately put into the ftores. And they are not to permit any goods either to be laden in, or landed from, any boat or fhip, without warrant from the collector. except fuch fine goods, or fmall parcels; and for thefe they are to have a furveyor's direction.

And they are to enter all their warrants, and alfo all the difcharges in a flock book; and to keep the warrants by themfelves until the fhip is difcharged; and then to deliver them to an officer, who in Dublin port is called the *Jerquer*; who was originally inftituted in order to bring the mafters and commanders of fhips to a due method of invoicing; and for this purpofe he is to compare the tide waiters bill, land waiters difcharges, and ware houfe note for goods remaining in the ftore, with the mafters invoice, and to place all down in a jerquing book; and to fee that they and the mafters entry do all agree; and if any difference be, to note it down, and to make a true return in his jerquing note to the commiffioners before fuch fhip be cleared, &c.

Jerquer, his duty.

There are alfo four Surveyors general of excife for the whole kingdom, one for each province, who are to vifit the

Surveyors general of excife, their duty.

the feveral diftricts of the feveral collectors given them in charge, as often as they poffibly can, and therein ftrictly to examine the feveral officers employed; to fee that they have acted properly, and with activity in every branch of their duty, and to direct and affift them in all particulars of their bufinefs.

C H A P. VI.

Of PRISAGE.

Prifage what, and how payable in England. PRISAGE is an ancient duty payable to the Crown by prefcription; and fignifies a certain quantity of wine taken for his Majefty's ufe, out of every fhip importing the fame.   And in England it is due at the rate of 1 tun out of 10; for which the Crown pays the merchant 20 fhillings, by way of compenfation for freight.

How in Ireland. But in·this kingdom the fettlement of prifage is as follows, viz. when the quantity of wines imported in any one fhip amounts to nine tuns, and under eighteen tuns, fingle prifage or one tun is taken; when fuch quantity amounts to or exceeds eighteen tuns, double prifage or two tuns are taken; but no money is paid to the merchant.

In kind or in money. It is either taken in kind, according to its original inftitution, or a certain fum is paid in lieu thereof by the importer.   If in kind, half the quantity is taken by the proper officers before the maft, and the other half from behind the maft.   When not taken in kind, the following rates are fettled, by agreement, to be paid in lieu thereof; and are received and accounted for by the collector of the

port

port where fuch prifage becomes due, in like manner as any other duties due to the Crown.

Rates fettled by agreement to be paid in lieu of prifage.

French wine, {For fingle prifage,   £30
            {For double prifage,   45
Malaga wines {For fingle prifage,   40
and Sherries, {For double prifage,   60
Canary,   {For fingle prifage,   50
            {For double prifage,   75

Prifage wines taken in kind are fet up to publick fale by inch of candle, and the produce thereof paid to the collector. There is alfo a cuftom of 15 fhillings per tun payable on prifage wines by prefcription; which cuftom is paid in lieu of all other duties whatfoever by the merchant importer, over and above the prifage or compofition for prifage, and not by the perfons to whom the prifage is due.

This duty was remitted or altered by Ed. I. in England, with regard to foreign merchants, by impofing a tribute of 2 fhillings on every tun of wine imported there, which was called *butlerage*; but this does not extend to this kingdom.

This duty was granted by King Henry II. in the year 1177 to Theobald the fon of Herveius Walter, to whom the King gave the butlerfhip of Ireland, whereby he and his fucceffors were to attend the Kings of England at their coronation, and prefent them with the firft cup of wine, for which they were to have certain pieces of the King's plate; and from thence it is faid the name of Butler is taken. And this duty was confirmed to

L 2   the

the faid family, afterwards earls, marquiffes and dukes
of Ormond, by feveral after grants; particularly in the
reigns of Edward III. Philip and Mary, and Car. II. and
became vefted in the late earl of Arran, by an act paffed
by the Britifh parliament in June 1721, enabling him to
purchafe the forfeited eftates of the late James duke of
Ormond, his brother.

Agreement
concerning
it between
the crown and
Ormond fa-
mily.

But in procefs of time, thefe cuftoms being extremely
troublefome in the collection of them, an agreement was
entered into between the Crown and the faid James late
duke of Ormond in the year 1704, whereby the faid duke
of Ormond was by deed or inftrument under his hand
and feal to empower the commiffioners of the revenue in
Ireland, by their officers, to collect and receive to her
Majefty's ufe and behoof the faid duties of butlerage and
prifage, for feven years from Michaelmas 1704; and in
confideration thereof, the yearly fum of £3500 was to be
paid to the faid duke of Ormond, his heirs, executors,
adminiftrators and affigns, out of her majefty's revenues
in this kingdom; which agreement was accordingly carried into execution, and her Majefty's letter and grant
accordingly had · and paffed for the faid annual fum.
Afterwards, the faid duke of Ormond executed a further
leafe to the then commiffioners of the revenue, on the
fame terms with the former, bearing date the 16th day of
Auguft 1707, for ten years and an half from Michaelmas
1711; and afterwards, by deed bearing date the 19th day ·
of November 1709, the faid duties were leafed for a further
term of ten years and an half, to commence 25th March
1722, for the like annual fum of £3500, which expired at
Michaelmas 1732.   And in the year 1733, the fame were
leafed by the faid Charles earl of Arran, for three years
from Michaelmas 1732, in confideration of the yearly fum

of

of £4000, for which the faid earl had the King's letter and grant as aforefaid; and they fo continued to be renewed for three years, for feveral years; but this being found very troublefome, the faid earl of Arran, in the year 1744, propofed either to collect thefe cuftoms himfelf, or to fet them for a long term, or at the will of both parties; which laft propofal was agreed on; and accordingly his late Majefty K. Geo. II. by his letter, dated at St. James's, 9th April 1744, appointed a yearly fum of £4000 to be paid to his lordfhip for the faid cuftoms during his Majefty's pleafure. And this agreement ftill fubfifts between the Crown and the heir of the faid earl.

It is faid the Crown is a confiderable lofer by the farming of thofe duties of prifage and butlerage; for that they don't amount annually to near the fum which the Crown fo pays for them.

A queftion has arifen, whether prifage wines, in the hands of a fubject, are liable to the duties impofed on wines by the act of excife 14 and 15 Car. II. c. 8. and the additional duties by fubfequent ftatutes; for, by the act of tunnage and poundage, prifage and butlerage are particularly excepted. This queftion depends on the conftruction of the act of excife, to which the other acts (by which the additional duties on wines imported are impofed) refer. By the firft claufe in the act of excife relating to wines a duty is impofed on all wines imported; fo that, if it had refted on that general claufe, there would have remained no doubt but that prifage wines would have been liable.

Whether prifage wines in the hands of a fubject are liable to the excife and additional duties.

But the difficulty arifes on the fubfequent claufe, by which all the faid duties are to be paid by the firft buyer,

before

before he receives the commodities from the merchant importer; or by the merchant importer, being a shopkeeper, retailer or confumptioner; and it may be faid that the grantee of the prifage is neither an importer, nor a buyer from the importer; and that this claufe explains the former, and confines the duty to fuch wines only as are bought from the merchant importer, or retailed or confumed by him.

However on confideration of the whole act and of the nature of prifage, it is holden by the beft opinions that the duty attaches immediately upon the importation of the wine; and that the latter claufe was not intended to difcharge the duty impofed by the former, but only regulates the collection, and the manner of payment of the duty in favour of the merchant.

And as this conftruction feems to fatisfy the words of the act, fo it is conceived to be reafonable both with regard to the publick and the grantee. The defign of the law makers (as appears from the preamble) was to eftablifh a certain revenue for the defence and prefervation of the realm; and therefore it fhould feem that the act is to be liberally conftrued for the benefit of the publick; and prifage being a cuftom due to the King for wines brought in by merchants, paying by prefcription 20 fhillings per tun, the grantee retains the full benefit of his grant againft the importer; fince his right of taking the prifage wines from the merchant at the prefcription price remains as it did, and is put on the fame foot with other fubjects, in cafe of his own confumption; but in cafe of fale, he is in no fort affected by the act, becaufe the duty falls on the buyer; whereas a contrary conftruction would leffen the fund defigned for the publick fervice, and tax the fubject

to

to enrich the grantee, which it is conceived could not have been the intention of the legiflature.

It is likewife apprehended that this conftruction is warranted by the authority of the judgment in the cafe of Paul and Shaw, in the Exchequer chamber in England, in Hillary term in the 8 Ann, 2 Salk. 617. where the queftion came to be, whether the grantee of the prifage in England was liable to the additional duty charged on wines by the 9 and 10 Will. 3. c. 4; in which cafe it was unanimoufly refolved that the grantee was liable. And it is not apprehended that the different penning of the Englifh and Irifh acts, in relation to the payment of the duty, will vary the cafe as to the prefent queftion; the Englifh act requiring the importer to give fecurity for the payment of the duty, and giving him the advantage of 10l. per cent. for prompt payment; and the Irifh act directing the payment to be made by the firft buyer; or by the importer, being a fhopkeeper, retailer, or confumptioner. And it muft be obferved that, though from the nature of prifage the grantee may in ftrictnefs be confidered as a buyer from the importer, yet he cannot in any refpect be deemed an importer.

However the duty, if any be due on the excife act or additional duties, has not been paid.

CHAP.

C H A P. VII.

L I G H T-H O U S E   D U T Y.

Light-houfe
duty, what.

LIGHT-HOUSE duty is a tribute of four pence per tun payable to his Majeftv, by his prerogative, by foreign fhips trading to Ireland, towards the fupport of his Majefty's light-houfes, which are erected here for the fafeguard of the lives of fea-faring men, and the prefervation of fhips and cargoes.

Granted to
Sir Robert
Reading in
truft for Lady
Mountrath.

King Charles the fecond, by letters patent in the 17th year of his reign, in confideration of fervices done by the Countefs Dowager of Mountrath, then married to Sir Robert Reading, did grant unto Sir Robert, in truft for Lady Mountrath, a duty of one penny per tun inwards, and one penny per tun outwards, to be levied on all fhips belonging to fubjects; two pence per tun in like manner upon all fhips belonging to ftrangers; ten fhillings yearly on fifhing boats; and upon all French fhips fuch a duty as Englifh fhips paid at Bourdeaux, provided that it fhould not be lefs than two pence per tun inwards and outwards; upon condition that he fhould build and maintain fix light-houfes in this kingdom.

Another
grant to the
Earl of Arran
under the
fame truft.

This patent was furrendered, and another granted in the 19th year of the fame reign to Richard Earl of Arran, for the term of fixty-one years, to the fame effect and on the fame truft.

Several

Several petitions were afterwards prefented to the Houfe of Commons of England, particularly from Chefter and Liverpool, complaining that the faid duties were a grievance and burden to trade; whereupon letters patent bearing date the 19th of July, 1672, were made out to Sir Robert Reading, granting him a yearly falary of £500 out of the concordatum money; and Sir Robert obliged himfelf by deed not to receive the duties payable by fubjects; but the duty on foreigners was ftill payable.

In the feffion of 1703, the Houfe of Commons of this kingdom, obferving this charge of £500, made inquiry into the execution of the covenants of the patent, (which was then become the property of the Earl of Abercorn by his marriage with Sir Robert Reading's daughter) and it appearing that only two of the fix light-houfes were kept up and thofe very ill fupplied and attended, they came to feveral refolutions which they ordered to be laid before his Grace the Duke of Ormond, then Lord Lieutenant. Thefe refolutions, in January following, were fent by his Grace to the commiffioners of his Majefty's revenue, with orders to make their report in relation to the fize and coft of two of the light-houfes; and in April, 1704, their excellencies the Lords Juftices gave the like orders in relation to the other four.

In confequence of the meafures taken upon the commiffioners report, the Earl of Abercorn furrendered his patent; and Queen Anne, by her letter dated 22d of November, 1704, entered at the fignet office, did direct the management of the faid light-houfes to be put under the care of the commiffioners of the revenue, and that the expenfes fhould be paid out of the revenue. But this letter gives the board no power to erect or maintain light-houses in any other places but there specified.

New light-house at Loophead.

In September, 1717, a memorial of the corporation, proteſtant merchants, and citizens of Limerick was preſented to the Houſe of Commons of this kingdom, upon which they came to a reſolution that the building a light-houſe on or near Loophead at the mouth of the river Shannon would be of extraordinary uſe to the publick, by preventing ſhipwrecks on the weſtern coaſts of this kingdom; which reſolution was laid before his Grace the Duke of Bolton, Lord Lieutenant, who referred all the papers to the commiſſioners of the revenue, with directions to determine the ſituation and expenſe of the intended light-houſe which was accordingly done; and his Majeſty King George the firſt, by his letter of the 25th of April, 1720, entered at the ſignet office, did order that the then commiſſioners of the revenue ſhould defray the charge of maintaining of the ſame out of the revenue at large. But neither doth this letter give any authority to the commiſſioners for erecting or maintaining any new light-houſes; and there are not any light-houſes now maintained out of the revenue but by virtue of theſe two authorities.

How the duty is paid.

The duty above-mentioned continues ſtill payable by all foreign ſhips trading to Ireland, and becomes due immediately upon their arrival in any port in Ireland. But no more than four pence per tun is taken for any one voyage, though ſeveral ports may be touched at in the courſe of it. The payment of it will appear from the receipt or certificate of the collector receiving the light-houſe money, which is given to prevent any diſputes that may poſſibly ariſe about it, on their putting into any other harbour in the kingdom during that voyage.

And

And this duty at prefent amounts to about £400 or ^{The amount} £500 yearly; and the expenfe of maintaining the feveral ^{of it.} light-houfes amounts to above double that fum.

C H A P.   VIII.

Of the INLAND EXCISE, ALE, &c. WINE, &c. LICENSES.

THE Inland Excise is the duty upon beer, ale, and Inland excife ftrong waters, granted to the Crown by 14 and 15 Car. 2. c. 8. after the following rates, viz.

For every 32 * gallons of ale and beer, of above fix On ale and fhillings the barrel price, brewed within this realm by the beer. common brewer, or in his veffels, or by any other perfons who fhall tap or fell out beer or ale, to be paid by the brewer, or fuch other perfons refpectively, two fhillings and fix pence.

For every 32 gallons of fix fhillings beer or ale, or under that price, brewed by the common brewer, or in his vef- fels, or by any other perfons who fhall tap and fell fuch beer or ale, to be paid by the brewer, or fuch other per- fons, fix pence.

For all aquavitæ or ftrong waters, diftilled within this On ftrong realm, whether of foreign or domeftick fpirits or materials, waters. to be afterwards vended, to be paid upon every gallon, by the firft maker or diftiller thereof, four pence.

* Which gallon is to contain 272 ¼ cubical inches.

M 2   Befides

Additional
duties.
Befides which perpetual duties there have been fince granted, and are now payable, the following additional duties, viz.

	l.	s.	d.
For every 32 gallons of beer or ale above fix fhillings price,	o	2	o
For every 32 gallons of beer or ale not above that price	o	o	4
For every gallon of aquavitæ, &c.	o	o	4

Brewers and
diftillers to
make weekly
entries.
And all brewers and diftillers liable to fuch excife are to make weekly entries on every Monday at the excife office, of the quality and quantity of all beer, ale, and ftrong waters brewed or diftilled the week before; and pay and clear the excife, on forfeiture of £20 for the firft week's neglect, £40 for the fecond, and £60 for the third; befides double the value of all liquors fo brewed or diftilled by them in fuch weeks.

Power given
the commif-
fioners to ap-
point gaugers.
And the commiffioners of excife have power given them to appoint fworn gaugers to enter, by night or by day, into any houfes, &c. belonging to any brewer or diftiller, and to gauge their veffels and take an account of their liquors. And the returns of fuch gaugers to the commiffioners or their fub commiffioner fhall be a charge upon the brewer or diftiller, if it exceeds the quantity by them entered.

Allowance to
common
brewers.
Common brewers, in paying their excife, are by the act to be allowed 64 in every 704 gallons of beer, and 32 in every 672 gallons of ale, and fo in proportion for a greater or leffer quantity by them brewed, free from all duties; which

which is to be deducted from their payments, in respect of filling, waste, leakage, returns or other accidents. But instead of the above allowances, pursuant to a letter from Lord Wharton, then Lord Lieutenant, to the commissioners, dated the 24 September 1709, an allowance of 2 gallons in 22 of ale, and 2½ in 23 of beer, is now given.

Here may be observed a material difference between the allowances in import and inland duties. In the first case, the deductions are always made out of the duty; but in the last, out of the quantity. In the first, the duty is always charged, and the £5, £6, or £10 per £100, is afterwards subtracted; or if goods are entitled to any abatement on account of damage received, the duty is always first paid down, and the allowance is given by way of drawback or repayment. In the last, the proportional quantity of liquor is deducted out of the total number of gallons, and no more charged than what duty is really paid for.

Difference between the allowance on the import and inland excise.

The gaugers likewise, on taking any gauge of warm wort, make an allowance of one tenth part, pursuant to an article in their instructions.

Allowance on warm wort.

And by 11 Geo. III. 2. and continued by 13 Geo. III. 2. a duty of one penny per gallon is to be paid out of all cyder which shall be sold or tapped out by retail.

The frauds and abuses practised by brewers and distillers require a more than ordinary circumspection, which has given occasion to numerous penal laws in relation to them too voluminous to insert here.

The

The officers for managing and collecting thofe duties are the gauger, furveyor, and collector of the diſtrict. The whole diſtrict is divided into walks, to each of which there is a gauger affigned.

Gauger, his duty. The gauger goes round his walk, twice every week; and takes an account of every brewing within it, and of the quantities of each fort of liquors made at each brewing in that compaſs; this he reduces into gallons, according to which the duty is charged.

Surveyor, his duty. The furveyor of the excife goes after the gauger, once in the month, and takes private notes, in his pocket book, of the feveral brewings in that month and the quantity and qualities brewed; which are compared with the gauger's book. And every month they both fign a return to the collector, which becomes a charge on each perfon therein mentioned, according to which the excife is received by the collector at his monthly office.

Duplicates of their returns fent to the commiſſio- ners monthly. Duplicates of thofe returns, figned by the collector, gauger, and furveyor, are fent monthly to the commiſſion- oners of excife. They are then examined by an exami- nator appointed for that purpofe, (called the examinator of the furveyor and gauger's books,) as to the computa- tion of gallons, and calculation of money. Whereupon the examinator makes out a charge againſt the collector, which he tranfmits to the Accomptant general.

The

The ALE and BEER LICENSES, and the WINE and STRONG WATER LICENSES form another branch of the hereditary revenue. The duties arifing from the former were granted by the 14 and 15 Car. II. 6. which enacts that none fhall keep any common ale houfe or tipling houfe, or ufe common felling of beer or ale by * retail without a licenfe; for which 20s is to be paid to the collector, for his Majefty's ufe, for every year that the perfon fhall be fo licenfed. And no fuch licenfe is to be granted for a longer term than a year from the Eafter preceding the date of it.

Ale and beer licenfes.

WINE and STRONG WATER LICENSES are founded on the 17 and 18 Car. II. 19. which enacts that no perfon fhall fell by retail any kind of wine, aquavitæ, ufquebaugh, brandy, or other diftilled ftrong waters, without a licenfe.

Wine, and ftrong water licenfes.

The rate of wine licenfes is fuch a fum as fhall be agreed upon, not lefs than 40s yearly in any cafe, nor exceeding £40 yearly within the city or county of the city of Dublin, nor exceeding £20 yearly in any other part of the kingdom.

Rate of wine licenfes.

The rate of ftrong water licenfes was by 17 and 18 Car. II. to be fuch fum as fhould be agreed upon, not lefs in any cafe than 40s, nor more than £10 yearly within

Rate of ftrong water licenfes.

* Which by 7 Geo. II. 3 is explained to be felling by any meafure lefs than a gallon.

the

the city or county of the city of Dublin, nor more than £5 yearly in other part of the kingdom. But by 3 Geo. III. 27. the rate of ſtrong water licenſes within the city of Dublin, and within four miles of the tholſel of ſaid city, was altered to ſuch ſum as ſhall be agreed on, ſo as none do pay leſs than £4, nor more than £10 yearly.

And no licenſe for ſelling wine or ſtrong waters can be granted for any term exceeding three years from the Michaelmas preceding the date of it.

By whom collected.
The power of granting ſuch licenſes, and appointing collectors of the duty, was by thoſe ſtatutes veſted in commiſſioners to be commiſſioned under the great ſeal in each county, and nominated by the Chief Governor or Governors of the kingdom out of the juſtices of the peace and others. But as no ſuch commiſſions were for many years paſt ſubſiſting, the duty was uſually collected by the collectors of exciſe; which power is confirmed to them by 33 Geo. II. 4. by which it is made lawful for the commiſſioners of exciſe or any three of them, and the collectors of the exciſe in their ſeveral diſtricts, to grant ſuch licenſes.

Cyder licenſes.
Beſides thoſe licenſes for ale and beer, wine and ſtrong waters, it is enacted by 11 Geo. III. 2 continued by 13 Geo. III. 2. for two years from the 25 December 1773, that no perſon ſhall ſell by retail any cyder without a licenſe, and that ſuch ſellers ſhall pay a duty of ten ſhillings yearly.

Theſe

These licenses and agreements are certified by the gauger and furveyor, to the commiffioners of excife, monthly, together with the inland excife, and are a charge on the collector for the duty.

(margin: Certified to the commiffi- oners of)

C H A P. IX.

Of HEARTHMONEY.

SO early as the conqueft, mention is made in domes- day book of fumage, (vulgarly called fmoke far- things,) which was paid by cuftom to the King for every chimney in the houfe.

(margin: Fumage.)

But its introduction into this kingdom was by the ftat. of * 14 and 15 Car. II. c. 17. and 17. and 18 Car. II. c. 18. by which a duty of 2 s. for each fire hearth &c. yearly, was granted to the Crown, in lieu of the court of wards, payable on every 10th of January, at one entire payment, by the occupier, and recoverable by diftrefs and fale of his goods.

(margin: Hearthmoney what.)

---

* About the fame period, viz. 13 and 14 Car. II. a like duty was granted by the Englifh legiflature to the Crown. But upon the revolution, hearthmoney was, by a ftatute of 1 W. and M. Eng. declared to be not only a great oppreffi- on to the poorer fort, but a badge of flavery upon the whole people, expofing every man's houfe to be entered by perfons unknown to him; and therefore, to erect a lafting monument of their Majeflies goodnefs in every houfe in the king- dom, this duty of hearthmoney was taken away and abolifhed. But this tax, which was created at a time when the proper fubjects of taxation were not fo well underftood as they are now, ftill remains a moft oppreffive burden on the occupiers of the wretched hovels in many parts of this kingdom.

Who exempt
from it.
And from this duty no perfon is exempt, except thofe who live upon alms, and are not able to get their livelihood by work; and except widows who fhall procure a certificate of two juftices of the peace in writing yearly, that the houfe which they inhabit is not of greater value than 8 s. by the year, and that they do not occupy lands of the value of 8 s. by the year, and that they have not goods or chattels of the value of £4.

Formerly
farmed by
counties.
This duty was formerly farmed yearly by counties at cant to the higheft bidder, who gave fecurity by bond for the payment of his rent, and collected the duty him-felf; and paid in his rent to the neighbouring collector of the diftrict, who was charged with his bonds.

Now collect-
ed by collec-
tors appoint-
ed by the
commiffioners
of the reve-
nue.
But fince the year 1704, it has not been farmed, but has been collected by collectors appointed for that purpofe, which appointment was by the 17 and 18 Car. II. 18. to be by the lord lieutenant or chief governors, and council. However they were conftantly appointed by the commif-fioners of the revenue, by virtue of a claufe in their patent for that purpofe. But fome doubts having arifen whether fuch appointment were ftrictly legal, by 3 Geo. III. 21. all fuch appointments were confirmed, and a power was thereby given to his Majefty, by commiffion under the great feal, to authorize and empower the commiffioners of the cuftoms and excife, or any three of them, to appoint fuch officers and collectors.

Who pay it
to the collec-
tors of the
diftrict.
And thefe collectors make returns of the number of hearths to the examinator of hearthmoney accounts, who is under the direction of the commiffioners of the revenue,
and

and from time to time pay their receipts to the collectors of the refpective diftricts in which their walks lie.

The collectors of the diftricts return the hearthmoney accounts with the other quarterly accounts to the commiffioners of the revenue, who fend them to the examinator of the hearth money, who returns yearly a certificate of the charge or produce of each collection to the Accountant general, who lays the credit part of the account before the commiffioners of the revenue for their approbation; after which it is brought into the general account of each collection, and depofited in the Auditor general's office, with the remainder of the account.

Who return thofe accounts to the commiffioners of the revenue.

The annual produce of this duty at prefent amounts to about £60,000.

C H A P. X.

Of other INLAND ADDITIONAL DUTIES.

BESIDES thofe perpetual inland duties there are now feveral additional temporary duties, which have been granted, from time to time, to the Crown; fome for particular purpofes to which they are appropriated by parliament, and others for the fupport of government generally.

The feveral inland temporary duties.

And thefe are upon coaches and other wheel carriages; gold and filver plate; cards and dice manufactured in the kingdom; hawkers and pedlars; and on ftamps.

N 2   And

And firſt, as to the duties on coaches and other wheel carriages, of which there are three.

Duty on coaches, &c. By the firſt, there is payable for 7 years, from 25th of March 1772, for every coach, chariot, &c. with 4 wheels, not uſed for hire, 20 ſhillings, to be paid yearly on the 10th day of January during the ſaid term; and for every chaiſe, chair, &c. with 2 wheels, not uſed for hire, 5 ſhillings yearly, payable as before. The produce of which duty is to be paid by the Vice treaſurer to the commiſſioners of the inland navigation, for the improvement of tillage and other uſeful purpoſes. 11 and 12 Geo. III. c. 4.

A ſecond. By the ſecond, there is payable for one coach, chariot, &c. with four wheels, which any perſon ſhall keep in his poſſeſſion, 10 ſhillings; and for every coach, chariot, &c. exceeding that number, which any perſon ſhall keep in his poſſeſſion, except hackney coaches and ſtage coaches, from 25th December 1773 to 25th December 1774, or from 25th December 1774 to 25th December 1775, 20 ſhillings, and for every chaiſe with 2 wheels, 10 ſhillings. But no perſon ſubject to the greater, is to be ſubject to the leſſer duty. 13 and 14 Geo. III. c. 1.

A third. By the third, there is payable for every coach, chariot, &c. with four wheels, which any perſon ſhall keep in his poſſeſſion, except hackney coaches and ſtage coaches, at any time during the ſaid years, 20 ſhillings. 13 and 14 Geo. III. c. 2.

How collected. And for the better collection of thoſe ſeveral duties, every perſon keeping ſuch coach, &c. is within a limited time to certify to the collector of the diſtrict an account

of·

of every coach, &c. which he shall keep; and the several duties are to be collected and levied by the collectors of the hearthmoney, or by such other persons as shall be appointed for that purpose, in the same manner as the duty of hearthmoney is collected and levied.

As to the duty on gold and silver plate, there is payable for seven years, from the 25th of March 1772, out of all gold and silver plate which shall be made or wrought in the kingdom, 6 pence for every ounce troy, to be paid by the makers and workers thereof. Which duty is to be paid to the commissioners of the inland navigation. 11 and 12 Geo. III. c. 4.

Duty on gold and silver plate manufactured here.

And no goldsmith or silversmith is to expose to sale any gold or silver plate, until it be assayed by the assay master; and if it be found conformable to the standard, then it is to be touched by the wardens of the company of goldsmiths, and marked with the marks used for that purpose; and then the said duty is to be paid by the person bringing it to be assayed and touched; and the assay master is, upon receipt of the duty, to stamp or mark it with such stamp or mark as the commissioners of the revenue shall from time to time appoint. ibid.

To be assayed and stamped, by the assay master.

And the assay master is to make entries, in a book, of the several quantities of plate so stamped or marked by him, and the duties received by him, &c. and once in every month to pay all the money received by him to the Vice treasurer. ibid.

Who is to receive the duty and pay it to the Vice treasurer.

As to the duties on cards and dice, there is payable during the said term of seven years from 25th of March 1772, for every pack of cards made in the kingdom 5 s.

Duties on cards and dice manufactured here

to

to be paid by the makers; which duties are to be paid to the commiffioners of the inland navigation. ibid.

A fecond on cards.   And there is payable a further additional duty of 6 pence for every pack of cards made in the kingdom, between the 25th of December 1773 and 25th of December 1775. 13 and 14 Geo. Iil. c. 1.

To be paid to the collectors of the ports.   And the makers of cards in Dublin, Cork, and Limerick, and of dice in Dublin and Cork, to which places refpectively the exercife of thofe trades is confined, are during thofe terms, once in every 14 days, to make a true entry upon oath with the collectors of thofe ports * refpectively, of all the cards and dice by them made within that time; and are once in 28 days to clear all duties owing by them, by paying the fame to the collector. 11 and 12 Geo. III. c. 4. 3 Geo. II. c. 12.

Cards and dice to be marked or ftamped.   And during the continuance of that term no cards or dice are to be fold, or expofed to fale, or played with, until fuch mark upon the dice, and fuch feal or ftamp upon the paper and thread enclofing every pack of cards, and fuch mark upon one of the cards of each pack, fhall be put, as the commiffioners of the revenue fhall appoint in writing, according to a power given them. 11 and 12 Geo. III. 4.

Duty on hawkers and pedlars.   As to the duties to be paid by hawkers and pedlars; there is to be paid by every hawker, pedlar, &c. (except perfons in the act excepted) travelling on foot or otherwife between 25th of March 1774 and 25th of March 1776, a duty of twenty fhillings by the year; and by every

* It fhould feem more agreeable to the nature of the duty to make the entry with the collector of the excife.

perfon

perfon fo travelling with an horfe or other beaft drawing
burden 20 s. by the year, over and above the faid firft
mentioned duty of 20 s. which duties are to be paid by
the Vice treafurer to the incorporated fociety for promoting Englifh proteftant fchools. 13 and 14 Geo. III. c.

And every pedlar fo travelling is, before the 25th of
March in each year, to deliver to the collector of excife
for the diftrict a note in writing how and in what manner he intends to travel; for which he fhall then pay the
duty; and thereupon obtain a licenfe from fuch collector
to travel or trade. ibid.

Licenfes to be obtained by them from the collectors.

As to the ftamp duties; they are a duty impofed upon
all parchments, vellum, and paper, whereon any legal
proceedings or private inftruments of almoft any nature
are written or engroffed, between the 25th of March
1774, and 25 of December 1775, upon all almanacks, news-
papers, advertifements, and pamphlets of certain forms
and fizes; and thefe impofts are very various, according
to the nature of the thing ftamped; from fix pounds to
one half-penny. 13 and 14 Geo. III.

Stamp duties.

For the better levying and collecting which duties
power is given to his Majefty, or the Lord Lieutenant
or Chief Governor of the kingdom, to nominate commiffioners or officers for ftamping and marking parchment, vellum, and paper, and managing the duties
thereon, who are to keep their head office in Dublin,
and they are empowered to appoint other inferior officers,
with the confent of the Lord Lieutenant, &c. for that
purpofe, and for the collecting and levying the duties.
ibid.

*Commiffio-
ners or offi-
cers to be ap-
pointed for
ftamping, &c.*

And

Parchment and paper to be carried to the office to be ſtamped.

And all vellum, parchment and paper, chargeable with the ſaid duties, is, before any of the matters or things in the act mentioned, ſhall be thereon engroſſed or written to be brought to the ſtamp office, to be ſtamped and marked; which the commiſſioners or officers are re-quired to do, on being paid the duty.   ibid.

No parch-ment &c. to be written on, till ſtamped, under penalty of £10.

And if any perſon ſhall engroſs or write upon any vellum, parchment, or paper, any of the matters or things for which the ſaid vellum, parchment, or paper, is chargeable with the ſaid duty, before it be ſtamped, or upon any vellum, parchment, or paper, that ſhall be marked for any lower duty than the legal duty, there ſhall be anſwered and paid his majeſty, for every ſuch deed, inſtrument or writing, the ſum of £10; and no ſuch record, deed, inſtrument, or writing, ſhall be pleaded or given in evidence in any court, or admitted to be good or available in law or equity, until as well the ſaid duty, as the ſaid ſum of £10, ſhall be firſt paid to his Majeſty's uſe, and a receipt produced for it under the hand of ſome of the officers appointed to receive the duties, and until the vellum, parchment, or paper, be ſtamped.   ibid.

How the duty is to be paid by the officers who collect it.

And all the officers concerned in the levying and col-lecting thoſe duties are to keep ſeparate and diſtinct ac-counts thereof.   And the perſons employed to collect and levy them in the city and county of Dublin are to pay the ſame into the treaſury on the firſt monday of every month.   And the perſons employed to collect and levy them in other parts of the kingdom are to pay them to the ſeveral collectors of the inland exciſe of the reſpec-tive

tive diftricts. And every ftamp officer, in fix days after his making any payment into the treafury, or to any of the faid collectors, is to give notice of the amount of fuch payments to the commiffioners of ftamps. ibid.

And none of the faid duties are to be received or collected by or paid to the faid commiffioners of ftamps; and the feveral perfons who fhall be employed in receiving, collecting, or paying the faid duties are, once in every year, to exhibit their accounts thereof to the commiffioners for taking impreft accounts, who are authorized to examine upon oath the faid perfons accountants concerning the money raifed or collected by them, and paid by them into the treafury or to the collector of inland excife; and they are to produce proper vouchers for any fum raifed received and paid. ibid.

Their accounts fubjected to the examination of the commiffioners of impreft accounts.

C H A P. XI.

Of SEIZURES, FORFEITURES, and FINES.

UNDER this head is comprehended all the revenue arifing to the crown, by its moiety or fhare of the produce of all feizures, which are condemned and fold as forfeited under the act of tunnage and poundage, and the act of import excife; as well as of fines impofed for breaches of thofe laws; to which the fubfequent acts creating the additional duties ufually refer.

The revenue from feizures &c. what.

Offences
against the
acts of custom
and excise
where triable.

And the offences committed againſt the former act are determinable in the court of Exchequer, by information or action, unleſs when otherwiſe directed by ſubſequent acts. But offences againſt the act of excise are tried before the commiſſioners or * ſub-commiſſioners appointed according to that act; from whoſe judgment there lies an appeal to the Lord Lieutenant, or other Chief Governors and privy council, or ſuch as they ſhall appoint under the great ſeal, who are called commiſſioners of appeals.

Goods con-
demned how
ſold.

And in the former caſe goods condemned are ſold by publick cant; in the latter by inch of candle. But the ancient courſe with regard to goods condemned in the court of Exchequer was, after they were appraiſed, to ſell them to ſuch as would give moſt above the value appraiſed at.

Regiſter of
ſeizures his
duty.

There is an officer called the clerk or regiſter of ſeizures in the port of Dublin, who takes an account of all ſeizures and of the produce of them. And when any ſeizure is returned to him, under the act of cuſtoms, by any ſeizing officer, he is forthwith to ſend a copy of the return (commonly called the ſeizing note) to the commiſſioners of the revenue, who direct their ſolicitor to prepare an information in the Exchequer; when under the act of excise, he ſends it to the clerk of the informations for Dublin port, who brings it to the ſolicitor, who prepares an information before the commiſſioners of excise.

* The ſub-commiſſioners are uſually the collectors of the diſtrict, the ſurveyors of the coaſt, port, and excise, and the Surveyor general of excise.

The

The regifter of the feizures receives the informer's moiety of all fines and feizures for breaches of the excife laws in Dublin port, for which he gives a particular receipt; and the King's moiety is afterwards paid to the collector, who gives a receipt for it. And he makes up and paffes his accounts quarterly with the commiffioners of the revenue, who examine it with the book of information of fales, which is attefted by the furveyor of the ftores and ftore-keeper. But in all other ports or diftricts the collectors receive the whole, and, after deducting the neceffary charges, divide the remainder between the King and in-former. The King's moiety of the money arifing from fines or the fales of feizures condemned in the court of Exche-quer is paid by the chief remembrancer, who fells fuch condemned goods, into the treafury; and the informer's moiety is in like manner paid by him to the informer. But the modern ufage is for the chief remembrancer to pay the whole to the commiffioners of the revenue, who diftribute it as above-mentioned.

To whom the produce is paid.

C H A P. XII.

Of the MANNER of PASSING COLLECTORS ACCOUNTS.

ALL the branches of the revenue hitherto treated of are accounted for by the collectors of the feveral diftricts in the kingdom in the following manner.

Collectors ac-counts when and where re-turned.

Every collector is to return weekly and monthly ab-ftracts, to the commiffioners of the excife and cuftoms, of his receipts and payments; and at the end of every quar-ter he is alfo to return, upon oath, a general account of the receipts and payments for that quarter; which quarterly accounts are to be compared and examined by the Accountant general, and the examiner and comptroller of the collectors accounts of incidents. And at the end of the year, to wit between the beginning of Easter and the end of Trinity term, the whole is to be drawn out ; and, when thus drawn out, he passes it first before the Accountant general, assisted by the clerk of the quit rents, where he is charged, as to the quit, &c. rents, by the rent roll on record in the Auditor general's office ; and he is to give an account of every particular sum received, for what land due, and from whom ; the residue makes up the charge. He also gives a particular account of the arrears in the same manner, with the reasons why they are not collected.

Passed before the Accountant general, and how charged, as to quit, &c rents.

As to customs and import excise. As to the customs and import excise, he is charged with the quarterly returns of entries made to the commissioners of the revenue, signed by the customer, comptroller, and himself.

As to inland excise, licenses, &c. As to the inland excise, and wine, &c. and ale, &c. licenses, he is charged with the monthly returns sent to the commissioners of the revenue, signed by the gauger, surveyor, and himself.

As to seizures. As to the seizures, his charge is the clerk of the seizures quarterly accounts, made up and compared by the commissioners of the revenue, and his receipts for the money arising from them.

As to hearth-money. As to the hearth-money, he is charged with the accounts of the collectors of the hearth-money, made up with the Accomptant general, and the vouchers for payment made to him by such collectors of hearth-money.

The

The difcharge of the collector of the diftrict on all thefe   How dif-
branches are Exchequer acquittances produced for payment   charged.
into the treafury; the arrears returned, which go in
charge to him for the enfuing year; the falaries to the
commiffioners and to other under officers; and all fuch
allowance as the commiffioners of the revenue are by their
commiffions empowered to make; as for fuits at law
where there is occafion, and the like.

The account thus ftated by the Accountant general is   The account
from him returned to the Auditor general with the original   returned by
vouchers and other matters relating to it.   ant general to
                                                           the Auditor
                                                           general.
The Auditor general re-examines and abftracts it under   Who exa-
general heads and engroffes it on parchment, and brings   mines and en-
the collector to fwear it before the Barons of the Exche- groffes it, &c.
quer, as is before mentioned.

The account thus engroffed and fworn is filed of record   And files it in
in the Auditor general's office, and the balance on the   his office
foot of the accounts is a charge on the collector in his next   when fworn.
account.

CHAP.

C H A P.　XIII.

Of FINES and AMERCIAMENTS, and FORFEITED RECOGNIZANCES; and their ESTREATS.

Cafual reve-
nue of what
confilling.

THE cafual revenue confifts of fines and forfeited recognizances, (commonly called the green wax) cuftodiam rents, profits of the hanaper, &c. together with fome other cafualties, as waifs, eftrays, felons and fugitives goods, &c. And as none of thefe are collected by the commiffioners of the revenue or their officers, but are all (except the profits of the hanaper) collected by the feveral fheriffs of the kingdom, and accounted for by them, and paid immediately into the treafury, it will be therefore proper to treat of them diftinctly. Befides the matters relative to thefe cafualties are chiefly tranf-acted in the court of Exchequer, where there are feveral rules and orders made concerning them in no fort relative to the other branches of the revenue.

Fines or
oblatas what
Madox 2~2.
Madox 273.

Madox 315.
Madox 320.
Madox 3??.
Madox 327.

FINES (anciently called *oblata* or offerings) and amer-ciaments made in the early ages a very confiderable part of the Crown revenue. The former were originally offerings or gifts to the Crown, for grants and confirmations of liberties and franchifes of fundry kinds; for liberty to hold or quit certain offices or bailiwicks; by tenants *in capite* for licenfes to marry, or that they might not be compelled to marry; for liberties relating to trade or merchandize; for the King's favour or good will, and that he would remit his anger and difpleafure; for the King's protection, aid

aid or mediation; to have feifin or reftitution of lands or chattels; and that perfons might not be diffeifed; to be difcharged out of prifon, and replevied or bailed to the cuftody of lawful men; for acquittals of various crimes, even homicide; and for a variety of other matter.

Madox 32.
332.
Madox 322.
Madox 341.
Madox 341.

But the moft remarkable head of this branch of the revenue was the fines paid to the crown for proceedings in the King's courts of juftice; as fines to have juftice and right; for writs, pleas, trials and judgments; for expedition pleas, trials, and judgments, or for delay thereof; fines payable out of the debts to be recovered.

On law proceedings.
Madox 293.
2.6.
Madox 308.
3.9.
Madox 311

Upon confideration of the nature of which feveral fines on law proceedings, it feems as if juftice or right was purchafed from the Crown. Againft which mifchiefs a remedy was provided by that claufe of *magna charta, nulli vendemus, nulli negabimus aut differemus rectum vel juftitiam*; which claufe feems to have its effect; for though fines for writs and procefs of law in many cafes were always a part of the Crown revenue, viz. from the time of the conqueft or foon after; and were conftantly paid after the making the great charter, as before, yet they were afterwards more moderate than they ufed to be before; and the actual denying of right and the ftopping and delaying of it, which before, upon paying of money or fines, ufed to be practifed, were by thofe charters quite taken away, or by degrees brought into difufe.

Formerly very oppreffive.
Madox 314.

So that of this great and indeed monftrous branch of the revenue, arifing from fines or *oblata* in this fenfe, all that remain are the duties arifing to the Crown for fealing patents and original writs, now ufually called the profits of the hanaper which are treated of hereafter; and poft fines

Now reduced to profits of the hanaper and poft fines

fines, fo called with refpect to the fines on the original
(or premier) fine, which are paid to the Crown on every
fine levied of land, *pro licentia concordandi* ; which are as
much as the premier fine, and half as much more.

The revenue arifing from AMERCIAMENTS or *mifericor-
dias* was anciently fo like that which arofe from fines or
oblatas, as to be fcarce diftinguifhable from it. But they
were generally fet for mifdemeanors or trefpaffes of different kinds; for diffeifins; for breach of affize; for
defaults or non appearances; falfe judgments; on hundreds
for murder or man flaughters, for not making hue and
cry, &c.

But the amerciaments moft neceffary to take notice of
here are thofe which were fet by the court of common pleas
in civil actions, either on the plaintiff, *pro falfo clamore*,
or on the defendant, for detaining a juft debt, by giving
judgment *quod fit in mifericordia*. And thefe judgments were
delivered to the clerk of affize, and by him to the coroner
of the county, who, according to the direction of *magna
charta*, c. 14. affeered or affeffed the amerciaments, and
afterwards delivered them back to the clerk of the affize;
and the judges eftreated them into the Exchequer.

But in procefs of time the coroners in all civil actions
kept one certain rule of amerciament, which became fo inconfiderable a fum as not to be worth the affeering; and
therefore they are now never levied.

Fines in criminal proceedings (which is the fenfe in
which they are ufually confidered at this day, and not in
that of *oblata* or offerings) were originally fet by the
King's bench and juftices in eyre, as ranfoms from imprifonment;

ment; and are ufually impofed by courts of juftice as a commutation of corporal punifhment for crimes and mif-demeanors; or for defaults or contempts of parties in fuits, jurors, &c. or for neglect of duty or mifbehaviour of officers of juftice.

All courts of record, where the fines are not granted away by letters patent, tranfmit the eftreats of them to the treafurer's remembrancer's office; and thefe, with the fines and amerciaments impofed in the Exchequer, were delivered to the clerk of the pipe formerly, who put them amongft the *nova oblata* on the great roll.

Eftreats of fines.

Another branch of the cafual revenue arifes from FORFEITED RECOGNIZANCES, which are bonds or obliga-tions of record acknowledged to the King, conditioned ufually for appearance at the court, to profecute felons, &c. to preferve the peace, &c.

Forfeited re-cognizances.

No recognizances were taken to the King by the ancient confervators of the peace, nor by the fheriffs or conftables; but, in cafes that were bailable, the fheriff or conftable took an obligation in his own name, but not any recognizance to the King. And the fheriff bailed to appear at his own torn, and the conftable to appear at the view of the frank-pledge.

Recognizan-ces not taken by confer-vators of the peace, &c

But when juftices of the peace were appointed, they iffued their warrant to apprehend the offender; and if it were a bailable offence, they by 3 Hen. VII. c. 3. Eng. bound him by recognizance, either to appear at the affizes or quarter feffions, and likewife bound over the evidence to profecute; and, if the offender or profecutor did not appear, the recognizance was forfeited, and the clerk of

By the juftices of the peace

VoL. I. P the

the feffions or of the peace refpectively eftreated it into the Exchequer.

Alfo recognizances taken for the King in his courts of record, or before juftices of the peace, are to be eftreated into the court of Exchequer when forfeited, that procefs may iffue on them.

But there are no recognizances eftreated out of the petty bag into the Exchequer, becaufe fuch recognizances, being for performing decrees of the court of Chancery, are taken to one of the mafters of the court, and not to the Crown; and therefore are fued there, and nothing is eftreated. And the ftatute ftaple and ftatute merchant are eftreated into chancery by the ftatute, and from thence execution is to go.

Fines with re-
gard to their
eftreats either
foreign.
Fines with regard to their eftreats may be confidered as of two forts; the firft are called foreign fines, viz. fuch as are impofed by the other fuperior courts of juftice, or by inferior courts, as the affizes and feffions, &c.

Or thofe im-
pofed by the
court of Ex-
chequer.
The fecond fort are thofe impofed by the court of Exchequer, in fuits commenced between parties there; as alfo on officers of juftice, &c. fome of which are in the pleas fide or in the chief remembrancer's fide, as on fheriffs, coroners, the purfuivant, &c. for not returning writs or procefs, or for not bringing in the bodies of perfons whom they have returned taken; others are impofed in the treafurer's remembrancer's fide; as on fheriffs for not accounting, not paying their lots into the treafury, or for not clearing their feveral accounts when ftated; as alfo for not returning writs; and on clerks of the Crown and peace, for not returning the eftreats of the affizes and feffions, &c.

And

And if thefe fines be not refpited or reduced, (or more properly difcharged) they are eftreated by the Exchequer, and the eftreats of them are made up by the refpective officers. And fuch as are on the pleas fide are to be delivered to the clerk of the eftreats or fummonifter, who is to iffue them in procefs to the feveral fheriff. But thofe in the Chief and fecond remembrancer's offices are to be delivered to the clerk of the pipe, in order to be written in charge and iffued in procefs. And thefe laft fines are of more confequence than thofe of the pleas or Chief remembrancer's office, and the parties on whom they are impofed, generally entitled to much lefs indulgence; and yet thefe fines are feldom levied, and little or no money is paid into the treafury for them, as they are generally reduced (or more properly taken off) on paying fome fmall fum to the poor box of the court. *How eftreated*

The eftreats from the court of common pleas are delivered to the Barons in open court, by the hands of the judges of that court, the laft day of each iffuable term. And in thofe eftreats are not only all fines and amerciaments in that court, but all poft fines on alienations and all the outlawries in that court; and thefe eftreats are by the Barons delivered over to the fecond remembrancer and by him fent down to the clerk of the eftreats; except the eftreats of the outlawries, which remain in the fecond remembrancer's office. *Eftreats from the common pleas.*

And the court of King's Bench, having both a civil and criminal fide. ought to deliver eftreats not only of fines fet on fheriffs and other officers, in the civil fide of the court, but alfo of fuch fines as have been fet on perfons *Eftreats from the King's bench*

P 2   for

for criminal offences. But from the acceffion of his late Majefty Geo. I. no eftreats were returned on either fide, (it is faid) for many years; and on both fides they are ftill on fome years unreturned. And in general eftreats are by no means returned as regularly and punctually as they ought to be, nowithftanding the following general rules and orders which have been made from time to time for the remedy of this grievance.

Rule 19th, February 1683.

Ordered that the clerks of the Crown and peace, and clerks of the markets, fhall enter their deputations with the clerks of the eftreats, and in the fecond remembrancer's office.

Rule 2d, December 1684.

Whereas it is obferved that the fheriffs, on their appofals, return many of the clerks of the Crown and peace to have no iffues to anfwer the fines on them for not returning their eftreats in time, whereby his Majefty is much damnified; for prevention thereof, it is ordered, that the clerk of the rolls do fend a lift of the names of the clerks of the Crown and peace to the fecond remembrancer's office, that purfuivants may iffue; and that their feveral deputies be required to enter their deputations in the faid fecond remembrancer's office, that it may be certainly known againft whom to iffue attachments.

Rule 28th, November 1692.

Ordered that the feveral clerks of the Crown and peace do enter their patents and deputations in the fummonifters office; and that Mr. Chetwood do give them notice having undertaken to do it; and that the Chief remembrancer do certify the names of fuch of the clerks of the Crown and peace as have entered their patents

with

with him, or their deputations with the fecond remembrancer and fummonifter, and a lift of thofe who have not entered them.

Ordered that all the clerks of the Crown and peace do return their eftreats to the fecond remembrancer, and that he deliver them to the clerk of the eftreats; and that after fo returning them the fecond remembrancer do forbear further fining; and that, when the clerks of the Crown and peace move to take off, the fines impofed before the returning the eftreat to the fecond remembrancer they produce a certificate from the fummonifter.

Rule 30th, June 1691,

The court taking notice that, for fome time paft, the feveral eftreats of fines forfeited recognizances, &c. returned into this court, have been firft lodged in the fummonifter's office, whereas regularly the fame ought, by the ancient method of the Exchequer, to be firft delivered into the fecond remembrancer's office, and entered there, and to be tranfmitted by the fecond remembrancer, to the fummonifter, and procefs to iffue from thence firft thereupon, and fo forwards to the other offices; it is thereupon ordered by the court that, for the future, all eftreats of fines and forfeitures whatfoever be returned into the fecond remembrancer's office and entered there, and from thence transferred to the fummonifter, who is to give a receipt for the fame, and to iffue procefs as ufual. And it is ordered that the fummonifter do not receive any eftreats, nor iffue any procefs thereon, until the fame are firft tranfmitted to him from the fecond remembrancer.

Rule 26th, April 1705,

The

Rule, 22d
November,
1703.
The Lord Chief Baron having taken notice of the inconveniencies that are occafioned by the neglect of the
clerks of the Crown and peace, and alfo of the commiffioners of oyer and terminer, in not returning the feveral
eftreats of fines impofed at the affizes and feffions in the
feveral counties in this kingdom; it is this day made a
ftanding rule of the court that all the commiffioners of
oyer and terminer, and the clerks of the Crown and peace
of the feveral counties of this kingdom, do return the
eftreats of all the fines impofed at the feveral affizes and
feffions into the fecond remembrancer's office, the next
term after the fame are impofed; that is to fay, the fines
of Lent affizes and feffions before the laft day of Eafter
term following; and the fines impofed at the fummer
affizes and feffions in Michaelmas term following; or in
default thereof the fecond remembrancer is ordered to
move the court the firft day of the following term to have
them fined for their neglect.

And in purfuance of this rule it appears that the protho-
notary and clerk of the Crown have been fined in the Exchequer for their neglect in not returning thefe eftreats.

Power given
the court to
amerce clerks
of affize, &c.
for neglecting
to return
eftreats.
And by ftat. 12 Geo. I. c 4. it is made lawful for the
Barons of the Exchequer to amerce clerks of affize, clerks
of the peace, or other perfons to whom it may belong to
make returns of eftreats into the faid court of Exchequer,
for neglecting or omitting to perform their duty in returning the faid eftreats, and to caufe the faid amerciaments to
be levied by fuch means as other amerciaments fet in the
faid court have been ufed to be done.

And

And the court will amend the eftreat of a fine in fome cafes, as was done in the cafe of the King againft John Henderfon; who being fined £30, for a mifdemeanor, by the juftices of the county of Down, which was afterwards reduced by them to £15, but by miftake of the clerk of the peace the whole was eftreated, it was ordered by the court that the procefs be amended according to the juftices order. 4th May, 1678.

Eftreate amended on account of the miftake of the clerk.

A man who lived within the liberty of the archbifhop of Canterbury was fined by the judges of oyer and terminer in Southwark for a mifdemeanor in court, which fine was eftreated; but no notice was taken in the eftreat of what place the man was; therefore Sir Conftantine Phipps moved that the eftreat might be amended by adding the place where the man lived, that the archbifhop who had the grant of the fines *tam integre tenentium quam non integre tenentium infra*, &c. might come before the foreign appofer and claim this fine by virtue of his grant; and faid that a man had been indicted and fined in Effex, which fine was eftreated here, and fuch an amendment made upon application; but to this it was faid, there was an addition in the indictment which was a guide to the court, being a record to amend the eftreat by; but here is a record for the King, and nothing but an affidavit on the other fide; and the court refufed to do any thing on the motion. June 26th, 1718, in the Exchequer in England. The King againft the archbifhop of Canterbury.

Or when there is a record to amend by. Bunb. 2.

So the party has been in feveral cafes admitted to plead or demur to an eftreat. As where a perfon pleaded to the eftreat of a recognizance, that the juftice of the peace did not take it in the county where he was a juftice; to which

Plea to an eftreat of a recognizance.

which the Attorney general demurred generally; and the defendant joined in demurrer. Mich. 1663, in the revenue side of the Exchequer here. The King againſt Mc. Cleary.

The like.

So where Oliver Keating, being fined, as clerk of the peace of the county of Longford, for not returning eſtreats of the ſeſſions, applied to the court for liberty to plead to the eſtreat, upon affidavit that another perſon, and not he, was clerk of the Crown and peace of ſaid county; and the court granted it. Michaelmas 1668, in the ſame court. The King againſt Oliver Keating.

Demurrer to an eſtreat of a fine, and rule for the Attorney general to join in demurrer.

So where the defendant demurred to a fine of £500 on him impoſed in the court of common pleas; and it appearing by affidavit that his Majeſty's Attorney general was attended with a copy of the demurrer, a day was appointed by the court for the Attorney general to join in demurrer. Same term, the King againſt Henry Nugent.

A plea to an eſtreat of a fine impoſed at a ſeſſions that the defendant was not within the juriſdiction of the ſeſſions.

Arthur Ward was fined for not attending as a jury-man at the general ſeſſions of the peace held for the city of Dublin at the tholſel of the ſaid city; which being eſtreated into the Exchequer, the ſaid Arthur Ward pleaded to the eſtreat that he was an inhabitant of St. Mary's abbey in the county, not within the juriſdiction of the court, and therefore not bound to attend. To this plea the Attorney general replied, that the ſaid Arthur Ward was, at the time of the ſaid fine, an inhabitant of Mary's abbey in the pariſh of St. Michan, ward of Oxmantown, and county of the city of Dublin, and within the juriſdiction of the court; and judgment and execution were afterwards had againſt the defendant for want of a rejoinder. Trin. 26 Car 2. King againſt Ward.

Mathew

Mathew Halley, clerk of the peace for the city and county of Londonderry, was fined the 16th of October, 1674, for not appearing and attending his office at the sessions of Newtownlimavaddy in the said county; which being estreated into the Exchequer, he pleaded a grant by letters patent of the office of clerk of the peace of the city and county of Londonderry; also a patent to the corporation of Derry to hold the session of the peace there and not elsewhere, and that it accordingly had been so held : that three justices of the peace issued a precept to the sheriffs of the city and county of Londonderry for a sessions at Derry, and that other justices had issued another precept for a sessions at Newtownlimavaddy, and that he attended at Derry. And upon the Attorney general's examining the plea and the patents; and it appearing to him therefrom, and from the estreats of the other justices, and affidavit of the clerk of the peace, that the several facts in the plea as set forth therein were true, he caused a *noli prosequi* to be entered on the record. Trin. 27 Car. 2. The King v. Mathew Halley.

Plea of matter of excuse to an estreat of a fine imposed at a session.

So, in the same year and term, the sheriffs of the city and county of Londonderry were in like manner fined by the same justices for not attending them as aforesaid, and like proceedings had.

And in the same term and year several fines to the amount of £2220 having been imposed on the sheriffs of the county of Londonderry, and others, at the session of the peace held at Newtownlimavaddy in the said county, it was ordered by the court that the estreats of the said fines should be staid from the file till further order; and on the 9th day of December following, on motion of his Majefty's Attorney general, they were ordered to be filed.

And on the 14th day of February, 1674, it was ordered that the faid fines fhould not iffue in procefs until the fecond day of the then next Eafter term, by which time the defendants were to plead thereto.

A like plea to an eftreat of a fine impofed at a feffion.

John Euftace and Maurice Euftace, clerks of the peace of the Queen's County, were fined by two of the juftices of the peace for the faid county, in a confidera- ble fum, for not attending a general feffions of the peace held in and for the faid county, on the 10th day of January 1675, to do their office; which being eftreated into the Exchequer, the faid John and Maurice pleaded to the faid eftreat, that they on that day attended the feffions held before other juftices; whereupon the like proceedings were had as in the former cafe Trin. 1675, the King againft John and Maurice Euftace.

So, in Hillary term in the fame year, John Sandes high fheriff of the faid county, was in like manner fined by the faid two juftices for not attending them; and the like pro- ceedings had.

Plea to an eftreat of a forfeited re- cognizance taken in K. B.

Thomas Roche having been bound in a recognizance in the King's bench in Michaelmas term, 1754, to appear in the fame term, and profecute Peter Hamilton in the faid court for divers charges and offences, and not to depart the court without licenfe; and having made default, in not appearing purfuant to his faid recognizance; and the fame having been therefore eftreated into the Exchequer; the defendant made fpecial application to the court for liberty to plead to the recognizance, and to the eftreat

and

and procefs grounded thereon, and that the faid procefs
might be ftaid; which motion came on in Hillary term
1755. And although the queftion then was whether
the defendant had a right to plead, and not whether the
plea he fhould plead was maintainable or not, yet the
latter was firft entered into: and it was urged on behalf
of the defendant, that the recognizance had not been forfeited, for that the defendant had appeared purfuant
thereto in the faid court of King's bench in Michaelmas term 1754, and was then and there ready, day after
day, to profecute the faid Peter Hamilton for the faid
charges and offences, but was not during the whole term
called upon for that purpofe; and that the recognizance
dropped for want of being continued; and that the court
could not continue the recognizance againft the confent
of the bail, nor extend the time in the condition; that
the fubject has a right to plead to fines and amerciaments, and by ftat. 5 Rich. II. Eng. to any debt due to
the crown; (by ftat. 33 Hen. VIII. Eng. he may plead
an equitable plea,) that the recognizance was not eftreated
truly or fully enough; that it was imperfect, in not faying when default was made; that fuch plea would not
be an averment againft the record, for that the eftreat is
not a record; that eftreats and procefs have been both
ftaid and amended; and that a *fcire facias* fhould have
firft iffued, to fhow caufe why execution fhould not
iffue, and it was compared to the cafe of fines in fenefchals courts; that there is no neceffity of fhowing to the
court what is intended to be pleaded; for when the plea
comes in, if it be frivolous, it may be fet afide, or demurred to; if not fo, it may be replied to; and Hard.
409, 471. Sav. 53. 2 Leon. 55. 1 Ld. Raym. 243 Lane
55. 4 Co. 71. Comb. 385, Maddox 367. 370. and the
cafes before mentioned were cited.

Q 2   On

On the fide of the Crown it was argued that the application was unprecedented; that none of the cafes quoted
came up to the prefent one; for they all were either
1ft, where there was a defect of jurifdiction in the court
below; or 2d, where fome matter of excufe *in pais* was
to be pleaded which was not an averment againft the
record; or 3d, cafes of fines impofed, which ftand upon
other principles.

1ft. The cafe of M'Cleary, in 1663, was a plea to the
eftreat that the juftice of the peace who took the recognizance did not take it in the county where he was
juftice; that this turned on its being no recognizance for
want of jurifdiction, and did not contradict the judges
certificate. The like of the cafe of the King againft
Ward, 26 Car. II. a fine impofed at the feffions of the
city of Dublin on an inhabitant of St. Mary's Abbey,
which was alleged not to be within the jurifdiction; and
the cafe in Hard. 471. was of a fine in a court leet for
breach of a by law.

That in the prefent cafe there was not any queftion of
the jurifdiction; the court eftreating, though called by the
other fide the court below, being the fupreme criminal
court of the kingdom; and that though this was mentioned to be like fines of fenefchals courts, and the
counfel for the defendant were for treating it with the
fame refpect only, yet the cafes were widely different;
the prefumption being in favour of fuperior courts; otherwife of inferior courts; and therefore it muft be taken
that the Kings bench had power to make the eftreat, and
have done rightly; and that their proceedings cannot be
reverfed by any court in this kingdom.

2dly,

2dly, That Savil 53. was a plea of a matter *in pais*, which was an excufe, and did not contradict or aver againft the record; and fo were all the precedents out of this court. That what was here defired was exprefsly an averment againft the record; which, though it may be as to inferior courts, cannot as to fuperior. Thus, to a prefentment delivered in a feffion and received, no averment lies that it was not affented to by 12; but it is otherwife of the prefentment of a court leet; for the party diftrained may aver that it was not prefented by 12. 1 Hawk. 130. 2d. ditto 162. Lib. affize 38. 21. Bro. abr. tit. record, pl. 45. a record of outlawry of divers perfons was certified into the Exchequer, among whom one was certified who was really not outlawed; and, on his goods being taken in procefs, he came, and faid that he was not outlawed; and parcel of the record came by writ of Chancery out of the court of King's bench into the Exchequer; and Green, one of the juftices of the King's bench, faid he was not outlawed, but that it was a mifprifion of the clerk. Shipwith (who was then chief Baron) faid " although all the juftices fhould record the reverfal they fhall not be believed, when we have the record that he is outlawed." 4 Co. 71. Hyne's cafe, one of the refolutions is that there can be no averment againft the record, though any may be taken which ftands with it; for the record fhall not be tried by *pais*.

That the eftreat, though perhaps not a record to all purpofes as in cafe of fines, for as to them it is only a minute of the judgment, Lane 55, (as where on a conviction of recufancy, *prout patet per eftreat* was pleaded in abatement, it was held ill, 1 L. Raymond 243.) yet, as to others, it is a record for the revenue; it is a record of the default

default below, the only record of it. That 4. Co. 71. fays
nothing of an eftreat. The fact of appearing or not can
be no other way tried than by the record; nor can any
fcire facias lie on the recognizance for this forfeiture even
below; fo that the council for the defendant muft miftake in faying they could have pleaded to a *fcire facias*
below. That this cafe of non appearance ftands on
another footing than other forfeitures, as it is a fact immediately within the knowledge of the court, and therefore requires no further information; confequently no
fcire facias neceffary, as where the forfeiture is occafioned
by an act not within their immediate knowledge; and that
the fame holds where an excufe is to be pleaded. So if the
recognizance be for appearance, and in the mean time to
be of good behaviour, this forfeiture muft appear on *fcire
facias*, or by conviction on an indictment, before it can
be eftreated. So where collateral matter of excufe is
pleaded, as death of principal, Savil 53. or his being
forceably taken away, Hunt's cafe Comb. 385. but that
thefe are not like the prefent cafe, for they all are confiftent with the record in admitting the non-appearance
and excufing it.

3dly, That the cafes of fines impofed are not applica-
ble, for the record is not contradicted by the plea, which
admits the impofition of the fine though it difputes the
caufe or authority of impofing it.

That the application was therefore *prima impreffionis,*
and its never having been made is an unanfwerable reafon
againft making it now. That of neceffity credit muft be
given to the relation of courts of their own acts; it is
the only proper proof and there never was an inftance of
a jury's trying it. That many judgments have been given

by

by default, and many writs of error brought on them, and fometimes errors in fact affigned; but it never was affigned for error that defendant did not make default, nor can it be affigned; for how can it be tried but by the record? and that has fet it out to be fo; fo that a *certiorari* would fignify nothing; but that there is a diftinction as to inferior courts. That Beaumont's cafe, 2 Leon 55. was thus; Note, it was holden by all the barons of the Exchequer, that a duty, which is not naturally a debt but by circumftances only, as debt upon a bond for performance of covenants or to fave harmlefs, may be affigned over to the Queen for a debt; but in fuch cafe a prefent extent fhall not iffue, but a *fcire facias* fhall iffue forth, to know if the party hath any thing to plead againft fuch affignment. But that was not like the prefent cafe; for what averment is there in it againft a record? the party before affignment would have had a right to plead; for the plaintiff muft have fued and declared; but that no fuit could have been on this eftreat below. That this diftinction of pleas was confiftent, and would rule and account for all the cafes.

That an application to the commiffioners of reducements is the ufual method to mitigate both fines and eftreated recognizances; but they never do it, till the parties ftand a trial; which was the true reafon of this extraordinary application.

As to the objection that the recognizance was not eftreated truly or fully enough as it does not alledge when the default was, it appears that the condition was to appear in Michaelmas term, when it was refpited to Hillary term, and the default was then; and as to what was faid of its having dropped for want of being continued,

the

the rule fhowed the fact to be otherwife: that if the party
be not difcharged the recognizance continues of courfe,
the condition being not to depart without licenfe. Farrefly
97. Owen and others of the city of Coventry were
bound by recognizance, and appeared for two terms,
and no profecution being againft them, it was moved
to difcharge their recognizance, or difpenfe with their
appearance; but the court faid they could not do it,
and that all they could do was to refpite their recognizance continued for more than a term. Regina verfus
Redpath. Fortefcue 358. Ca. law and equity, 152. And
that as to the objection that the court could not continue
the recognizance againft the confent of bail, that is true,
if bail deliver the principal in court and defire to be
difcharged, which was not this cafe; elfe the court may
continue them. That poffibly the defendant might have
applied for a pardon, and that therefore, fince fome things
may be pleaded and others cannot, a previous application
for leave to plead and ftop the procefs becomes neceffary;
and it muft on fuch application be fhown what plea is
intended, as on leave to plead double matter; but that the
plea mentioned by the counfel for the defendant being a
direct averment againft the record ought not to be received.

That as to what was infifted on, that it need not be
fhown to the court what is intended to be pleaded; for
that when the plea comes in, if frivolous, it may be fet
afide or demurred to; if not, it may be replied to; in
anfwer thereto, the application for leave proves the
granting it difcretionary, and the ftopping the procefs is
confeffedly fo, and therefore the difcretion of the court
muft be determined by the confideration whether the
party be entitled to relief; and this neceffarily obliges
him to inform the court what his excufe or cafe is, or, in
other

other words, what plea he intends to put in. If the plea, being true, will be a good one, he ought to be admitted to plead; but if the plea offered be such as by law he cannot be admitted to prove, (which was the present case) or, being true, is no foundation or cause to relieve, the court in either of these cases ought not to receive the plea or stop the procefs; and that even fuppofing fuch fham plea fhould be admitted, it is not a confequence that the procefs fhould be ftopped.

That if this be a matter of favour the party fhould be in court to afk it. Where parties are convicted the court never fuffers motions in arreft of judgment or for new trial but in the prifoner's prefence, in order that if the motion goes againft him the court may have him in their power.

Upon the whole, the court were of opinion that the plea ought to be laid before the court; which was accordingly done, and ferved on the folicitor for the Crown; to whereupon, and on hearing counfel on both fides, and it being admitted that pleas in fuch cafes had been received, the court were of opinion it was reafonable in this cafe, and accordingly ordered it to be received, and that all things fhould ftop till further order.

The plea was, that the defendant had appeared in the faid court of King's bench purfuant to the condition of his faid recognizance, and was then and there ready to profecute the faid Peter Hamilton as he was alfo bound to do. It was then confidered whether the plea was to be replied or demurred to ; and upon confulting Sir Robert Henley then Attorney general of England thereon, he was of opinion that the plea was bad and fhould be demurred to; that it tended to falfify the record or (which is more abfurd) to join iffue upon a matter in law to be tried by the country; for that it feemed to be intended to try the effect of the recognizance by a jury. That if the eftreat was irregular it fhould have been fet right by an application to the court; but that it feemed to be regular, and upon the whole of the cafe the recognizance to be forfeited.

Accordingly a general demurrer was afterwards filed, and the plea with leave of the court was afterwards amended; but the defendant having confented to judgment, and applied for a *nolli profequi*, and the Attorney general here being made acquainted therewith, and confenting to the procefs being ftayed, the fame was ftayed; and in Trinity term 1770, an order was conceived on a motion made by the Attorney general for the defendants, on the warrant of the Lord Lieutenant that the fame fhould be received, and that fatisfaction fhould be entered on the record of the judgment againft the defendant and his bail, and that their recognizances fhould be difcharged purfuant to the faid order. The King againft Roche.

May be reduced either before or after they are eftreated.

But all thefe fines, &c. as well foreign, as thofe impofed by the Exchequer in fuits commenced in either fide of the court between party and party as aforefaid, may be reduced as well before as after they are eftreated.

Reducements by commiffioners of reducements.

But if the foregoing fines cannot be reduced by the court of Exchequer, as the court of Exchequer, they may be reduced by the commiffioners of reducement, who are the Lord High Treafurer, Chancellor and Barons of the Exchequer, in prefence of one of the

King's

King's council, and are appointed by commiffion under the great feal; and upon a petition to them, they at their difcretion reduce the faid fines, generally to a very fmall fum, often to fix pence; which being paid into the treafury, and the feveral officers being fatisfied their fees, the party is no further troubled or molefted *.

And the fines impofed by the court of Exchequer are entirely under their own power; and are either refpited or reduced (more properly difcharged) on confents from the attornies concerned, and upon motion thereon. But if they are reduced (or rather difcharged) it is upon paying fome acknowledgment into the poor box of the court; which was originally intended for the ufe of the poor, but afterwards came to be equally divided among the Barons, except fuch part as they thought proper to give among the poor. And originally the court ufed to barter with the attornies for they fums the fhould pay on difcharging thefe fines; which feldom exceeded a piftole, or at moft two or three, be the fum to which fuch fine amounted ever fo great. But in the year 1716, the court thinking it below their dignity thus to barter with the attornies for the poor box money, a rule or declaration was made (as it is before faid, tho' no fuch rule is to be found in the fecond remembrancer's office,) that none of thefe fines fhould be reduced for the future unlefs

Fines impofed by the Exche-quer, how reduced.

---

* In the cafe of the King againft Thomas, Eafter term 1752, the Chief Baron faid that where a fine is properly impofed, but there are fufficient reafons for reducing it, this muft be done by petition to the commiffioners of reducements; but that where there is a miftake in the eftreating of any fum either by fine or recognizance, the court, as the court of Exchequer, may upon motion and affidavit of the facts, and without any petition, order it to be difcharged; and this order is to be entered on the roll.

fixpence

fixpence per pound were paid to the poor box for the firft hundred pounds, and threepence per pound for every other hundred pounds, of fuch fines. But in a committee of the Houfe of Commons of this kingdom, on the 6th day of November 1723, it was refolved that this rule was obtained from the court of Exchequer by furprife and was a grievance to the fubject; fince which time the poor box money on fuch reducement is paid much in the fame manner as it was originally and before the faid rule of court was made.

No part of them paid into the treafury, unlefs levied by the Sheriff.

But no part of thefe fines is ever paid into the treafury, except they are iffued in procefs and levied by the fheriffs, which (as is faid before) very feldom happens; and even in this cafe, the clerk of the pipe feldom *debets* them, but lets them remain a continued charge on the fheriffs. But the court of Exchequer may reduce or difcharge them, even after they are eftreated by them, upon confent of attorney and upon motion thereon, or for fuch other reafon as they fhall think meet.

Often kept on foot by management.

But thefe laft mentioned fines are often kept on foot for years, by the management of attornies in procuring confents for refpites; which refpites are often kept by the attornies, and revived from time to time, as occafion requires; and, it is feared, have been fometimes fictitious. Thefe improper proceedings not only take up much of the time of the court moft unneceffarily, but are alfo productive of great mifchief and inconvenience to the publick by the delay and failure of juftice, which muft of courfe be the confequence. Befides they often have been very prejudicial to high fheriffs, who are generally
ftrangers

ftrangers to thefe refpites until the fecurities for their fub-fheriffs, by management between the fub-fheriffs and the attornies of the courts, become infufficient; therefore when thefe fines amount to forty pounds, or fome certain fum, they fhould never be refpited further than the next fitting of the court, and fhould then be taken off or abfolutely eftreated.

Ordered by the lord chief Baron and the reft of the Barons, that all and every perfon and perfons that fhall hereafter obtain any rule either for refpite or difcharge, fhall take out their orders the fame term, or at the fartheft before the laft of the eight days after every term, and enter the fame with the officers where the faid debts are in charge; otherwife no orders fhall be drawn thereon; except upon further motion the court fhall give order for the fame.

Ordered that all parties do take out their orders of reducement, and pay in their money in fix weeks, or lofe the benefit thereof.

Ordered that in all orders of difcharge, or orders of refpite, there be inferted a claufe, that the faid orders be entered with his Majefty's commiffioners of the revenue, without paying any fee for the fame, or they to give out any copy thereof to the prejudice of the treafurer's remembrancer's office.

Ordered that all perfons who fhall obtain any order for refpite, reducement, or difcharge of any charges which have iffued in procefs againft them be obliged to profecute and pay fuch reducements the fame term fuch rules are obtained, or within eight days after; and that fuch rules

Rule 21 Nov. 1684.

Rule 8 Dec. 1680.

Rule 17 Dec. 1684.

Rule 5 Mar. 1684.

as

as are obtained after term be profecuted as aforefaid by
the laſt day of the term following, or to have no benefit
of the faid rule; whereof all officers and perfons concerned
are to take notice.

Rule 28 Nov.   Ordered by the court that for the future no fines
1752.   impofed on any ſheriff's for not returning writs direſted
to them ſhall be reduced, without producing an affidavit
affigning the caufe wherefore they delayed returning the
fame; and this to be a ſtanding rule on every ſide of the
court.

Rule 24 July   Ordered that all confents and affidavits for the refpiting
1772.   of fines impofed by this court upon ſheriffs, officers, and
miniſters of the court, and others, for their negleſts and
defaults, ſhall be filed in the fecond remembrancer's office
on or before the day but one next preceding the firſt revenue day after every iſſuable term.

This branch of the cafual revenue is of much more
confequence to the publick than it is generally underſtood
to be; and if more attention was, than is at prefent, given
to the management of it in its feveral ſtages, by thofe
who are concerned therein, it would, befides the increafe
of this branch of the cafual revenue, much contribute to
promote that due execution of the laws which is fo much
wanted in this kingdom, and to the preventing the many
riots, bloodſheds, and murders, for which it is at prefent
noted; but as in many other inſtances it happens that the
wifeſt regulations have been fruſtrated and rendered nugatory by the negleſt of fome fubordinate fpring, which in
the grand machine feemed fcarcely worth attending to,
fo it is in regard to thefe fines and recognizances; for it
too often happens, from want of due attention to the
latter,

latter, that neither the addition or places of abode of the parties who are bound to profecute offenders, are inferted therein ; or that if they are, they are omitted in the eftreats ; or that perfons of no property, credit, or repute, and often perfons under fictitious names are taken as bail for the moft atrocious offenders; fo that procefs is iffued againft them, at great expenfe to the Crown, to no pur-pofe, and the publick juftice of the kingdom, in this moft effential branch of it, either abfolutely defeated or greatly obftructed.

Now the prevention of this mifchief is much in the power, and indeed is a part of the duty of the juftices of the peace, by being careful to afcertain the profecutors of offenders brought before them, by their additions and places of abode, and binding them in a fufficient fum to appear and profecute ; and by taking due care not to accept any perfons as bail without full knowledge of their credit and fufficiency. And it is alfo to be wifhed that the fame caution and precifion were ufed, whenever a perfon is fined for an offence by a court of juftice.

And for the enforcing of this a late rule has been made, viz.

Ordered, upon motion of Mr. Attorney general, that the feveral clerks of the Crown and peace of the kingdom, do for the future infert in their eftreats the additions and places of abode of the feveral perfons mentioned therein, who have either been fined or have forfeited recog-nizances. And that when fuch additions or places of abode have not been mentioned in the recognizances which have been taken by the juftices of the peace, they do fo mention it in their returns, with the names of the

Rule 2 June 1772.

the juſtices who took the recognizances. And that on all
fines hereafter to be impoſed in any of his Majeſty's
courts of record in Dublin or elſewhere, commiſſions of
oyer and terminer, as alſo at the aſſizes and feſſions, and
other courts where fines or amerciaments are uſually laid or
impoſed, the ſeveral clerks of the Crown and peace, or
other proper officers, do immediately enter down the
additions and places of abode of ſuch perſon or perſons ſo
fined, and return the ſame in their eſtreats; and that the
Solicitor for the caſual revenue do cauſe this order
to be ſerved on the ſeveral clerks of the Crown and peace
of the kingdom or their deputies.

C H A P.   XIV.

Of the PROCESS which ISSUES to the SE-VERAL SHERIFFS called the PROCESS of GREEN WAX.

THE method of iſſuing this proceſs out of the ſummoniſter's office, the pipe office, and treaſurer's remembrancer's office, all which are uſually called the green
wax proceſs, is as follows, viz.

Summoniſ-
ter's proceſs.   The clerk of the eſtreats and ſummoniſter iſſues in proceſs twice every year, viz. in Trinity and Hillary vacations, all fines and amerciaments, forfeited recognizances,
&c. which are eſtreated and returned into the office from
the courts of King's bench and Common pleas, and from
the ſeveral clerks of the Crown and peace, which firſt proceſs (ſometimes particularly called the green wax proceſs)
is againſt the goods only, and is returned by the ſheriffs
yearly,

yearly, when they come on their accounts; and, after they
have compared with the fummonifter, they bring them to
the tranfcriptor and foreign appofer, who thereon appofes
the fheriffs in court on their accounts.

And the foreign appofer makes out a tranfcript in parchment of all the fums for which the fheriffs do not * tot,
which are called ‖ nils, which he fends down to the clerk
of the pipe, and fends a tranfcript or copy thereof to the
comptroller of the pipe, who, in the vacation of the iffuable
term after the appofal of the fheriff, fends them in procefs
under the feal of the Exchequer to the feveral fheriffs;
which procefs is called the § fummons of the pipe, and is
againft body, goods, and lands; and upon the fheriffs
accounts he appofes the fheriff in court thereon.

Procefs of the pipe.

And fuch fums as the fheriffs are not thereon charged
with, which are called nils, as aforefaid, if they are not
reduced or difcharged by order, the clerk of the pipe
tranfcribes out of the great roll into what is called a
paper book and fends them to the comptroller of the pipe,

*Second pro-
cefs of the
pipe.*

---

* *Tot,* i. e. *Totum in manibus.*   ‖ *Nils,* or *nihil:* i. e. *nihil in manibus.*

§ Lord Chief Baron Gilbert, in his Treatife of the Exchequer, page 133, fays
that the procefs of the pipe is certainly an unneceffary procefs, and fpends a great
deal of time to no purpofe; fince thefe fums have been already in charge by the firft
procefs, and coming out of that office nihill'd, that would have been fufficient authority for the clerk of the pipe to tranfmit them in fchedula pipæ; for magna charta
is fatisfied, fince it appears on the faid firft procefs that they had no goods or chattels; and therefore to iffue the fummons of the pipe is unneceffary. But however
that may be the cafe in England, it feems not to hold here; for the procefs of the
pipe there is againft goods and chattels only, and in the nature of a feri facias;
whereas in this kingdom it is againft goods, chattels, body, and lands: fo that in
fact it is the fubfequent procefs, viz. the fecond remembrancer's, that feems unneceffary; at leaft where the debtor himfelf is living.

who, in the Trinity vacation next following, again iſſues them in in § proceſs to the ſheriff.

Second re-
membrancer's
proceſs.

And what are *nil'd* in this proceſs the clerk of the pipe alſo makes a ſchedule of, which is called the ſchedule of the pipe, and ſends it once a year to the treaſurer's re-membrancer, who, every Trinity vacation, iſſues the ſame in proceſs, which is called the ‡ treaſurer's remembrancer's proceſs, and is againſt body, goods, lands, heirs, executors, and adminiſtrators. And what ſums the ſheriff *tots* for upon this proceſs are by the ſecond remembrancer certified to the Auditor general, who draws a tranſcript thereof, which is called the ſheriff's * foreign account, and ſends it to the pipe for the purpoſe of making out the *debets*.

Which uſed
to be conti-
nued.

And what was not totted for by the ſheriff in this laſt proceſs uſed to be renewed and continued to be iſſued in proceſs by the treaſurer's remembrancer until the debts

---

§ This ſecond proceſs of the pipe ſeems to be a moſt unneceſſary, ſuperfluous proceſs, and is directly in oppoſition to the rules of 25th November, 1685, and of 28th November, 1709.

‡ This is alſo called the long writ or prerogative writ, and is not iſſued till a *nihil* is returned upon the ſummons of the pipe; which was ſettled to be a part of the liberty of the ſubject by *magna charta c.* 8. *nos vero vel ballivi noſtri non ſeiſiemus terram aliquam vel redditum pro debito aliquo, quamdiu catalla debitoris præſentia ſufficiunt ad debitum reddendum, et debitor ipſe paratus ſit inde ſatisfacere.* Gilb. Treat. of Exch. 123. And yet as the ſheriff is bound to hold an inquiſition on this writ, whether the debtor had any goods and chattels, before he extend the lands, or take the body of the debtor, it ſhould ſeem as if this writ might be uſed in the firſt inſtance without any violation of *magna charta*; and thereby the perſons againſt whom theſe proceſs iſſue prevented from making fraudulent ſales or removing themſelves and their goods into other counties.

* So called becauſe it is made up from matters not in the pipe, ſuch as the chief remembrancer's *conſtat* of 15 ſh. for fines for *profers*, the foreign appoſer's *conſtat*, and the ſheriff's certificate of waifs and eſtrays, &c.

werc

were paid or difcharged by pardon or by reducement, (which may be at any time before the money is actually paid into the treafury;) or until he was otherwife directed by the court. But this having been attended with great expenfe to the crown, a rule was made the 28th of November, 1709. (which fee hereafter) that this procefs fhould iffue but once unlefs by particular order.

And note that all thefe procefs and tranfcripts are iffued by the refpective officers *ex officio*, without any fee or fees being ever paid by the Crown to any of them for making out or iffuing the fame, other than the ancient fees due to them on their patents or the eftablifhment, and allowed by the Crown; to wit, to the clerk of the pipe 55l. annually, to the comptroller of the pipe 7l. to the fecond remembrancer 7l. 15s. 6d. and to the foreign appofer 15l.

Thefe feveral procefs iffued ex officio.

Hence it plainly appears that upon the firft procefs of the green wax the fheriff muft either *tot* or *nil* according as the cafe is and cannot § *o'ni*; for the whole account muft appear upon the pipe roll that the clerk of the pipe may iffue *debets* for the payment of the fums fo *totted* for into the treafury; unlefs they be before difcharged by order of the Exchequer. Befides, many of the fums in this procefs being fmall and paid upon the firft demand, they were part of an annual charge in the fame manner as the other annual revenue of the King was. Nor can he *o'ni* on the fecond remembrancer's procefs as the procefs ends there; and as all that is poffible has been done to get them in by holding an inquiry upon the prerogative writ. But he may *o'ni* in the comptroller of the pipe's procefs as this is a tranfcript or duplicate of the pipe procefs.

Sheriff cannot o'ni on the firft procefs or on the fecond remembrancer's procefs.

§ *O'ni, i. e. Oneretur nifi habeat fufficientem exonerationem.*

S 2   And

Sums o'ni'd for continued in charge against the sherif.

And the fums fo *o'ni'd* for therein are continued in charge againſt the ſheriff until he clears his accounts; which if he neglects to do, fines are impoſed on him in manner as is hereafter mentioned. But a *tot* in the ſecond remembrancer's proceſs, as well as in the fummoniſter's proceſs, for any fine that is reducible, is confidered as an *o'ni.*

Rule 25th November, 1635.

Ordered for the future that none of the officers of this court do iſſue proceſs but once ; and after that draw them down to the ſecond remembrancer, to the end the prerogative writ may iſſue where the ſheriffs on their appoſal return neither body, goods, or lands.

Rule 22d February, 1685.

Memorandum, it is this day ordered that the proceſs to be iſſued for the King be made returnable the third return of Trinity term ; the court confidering the inconveniency which attended the refpective ſheriffs returning a *Tardè*; and to prevent their having any pretence that the proceſs came late to their hands.

Rule 25th February, 1695.

The court this day taking notice that the ſeveral ſheriffs of this kingdom do from time to time make very ill and infufficient returns on the proceſs of the pipe, do hereby order that every ſheriff of this kingdom ſhall for the future call a jury for their better information on their ſaid proceſs of the pipe; and that the comptroller of the pipe do for the future infert a memorandum at the bottom of their proceſs, requiring each ſheriff of the kingdom to be informed as aforeſaid.

Rule 28th November, 1719.

Mr. Solicitor general on behalf of her Majeſty informs the court, that the proceſs of green wax iſſuing firſt againſt
the

the goods from the fummonifter's office, and on the *nils* returned by the refpective fheriffs on that procefs, the feveral fums charged in that procefs and fo *nill'd*, are by the fummonifter transferred to the pipe, and from the pipe are iffued in procefs againft body, goods, and lands; and all the refpective charges which on that procefs are *nill'd* by the refpective fheriffs on their accounts are therein drawn down to the fecond remembrancer's office, and that thereon the fecond remembrancer iffues forth procefs againft bodies, goods, lands, heirs, executors and adminiftrators of the refpective perfons, for the refpective charges fo drawn down from the pipe office to the fecond remembrancer's office; from whence the procefs iffues for feveral years, though the refpective fheriffs, both by the inquifition held and returned on that procefs and likewife on their account, on their oaths, do return that there are not fuch perfons, goods, lands, heirs, executors or adminiftrators to be found in their refpective bailiwicks; and that procefs continuing to iffue becomes very voluminous and chargeable to her Majefty whereas really her Majefty derives no advantage thereby; and therefore on behalf of her Majefty prays that for the future procefs fhall not iffue more than once out of any office againft goods, bodies, lands, heirs, executors or adminiftrators of any of the perfons charged in the faid refpective procefs; and that on the return of the refpective fheriffs on the procefs out of the fecond remembrancer's office, an exannual roll may be made up to lie by, and not be iffued in procefs, but by particular order of the court, againft any of the perfons, their heirs, executors or adminiftrators; and that commiffions do iffue, when the court fhall think fit, to commiffioners to be appointed by the court on faid exannual roll,

to

to inquire and find out what may be had or levied thereon; court ordered accordingly *.

Rule 14th May, 1717.

All fheriffs fhall hold inquiries on the fecond remembrancer's procefs in every barony in their refpective counties at their peril.

All proceffes for fines eftreated to be delivered to the purfuivant and by him to the fheriffs.

The feveral proceffes that iffue for all fines eftreated, both foreign fines and fines impofed by the court of Exchequer, are to be delivered to the purfuivant, who is to deliver them to the feveral fheriffs of the kingdom.

And the time and manner of delivering thefe proceffes we find fettled by the following rule:

Rule 4th February, 1709.

The court being informed by Mr. Attorney General Forfter of the great delays in delivering the feveral proceffes of green wax iffuing out of the court to the feveral fheriffs of this kingdom, which is of great coft and prejudice to her Majefty; the faid fheriffs not having fufficient time to execute the fame; It is ordered, that all fuch proceffes be delivered to the purfuivant, in three weeks after the end of every iffuable term, by the feveral officers of this court iffuing the fame; and that the purfuivant do deliver them to the feveral fheriffs of this kingdom in three weeks after the delivery of them to him.

---

* Notwithflanding the above rule, yet where a fheriff neglects to account, in order that procefs may not be wanting in the feveral departments feveral names and fums are taken by the fecond remembrancer out of the paper book for the year before, for the comptroller of the pipe's procefs, and two of the roll before the paper book which is called by fome the exannual roll for the fecond remembrancer's office.

And

And thefe fines, &c. uncollected are, as has been before faid, continued in the great roll, and iffued yearly to the fherifis in procefs, until fuch time as the court of Exchequer fhall think fit to ftrike them out of the annual charge and place them in the exannual rolls as defperate debts; but before that is done, a commiffion ought to iffue, directed to difcreet men, to inquire and return on their oaths whether any thing is to be gotten of thefe debts. But however it is generally practifed otherwife; for when thefe fines, &c. have been in all the procefs in the manner before mentioned, fo that there appears no likelihood of getting any thing, the clerk of his own accord leaves them out, but lays them up carefully in his office as defperate debts; yet they may at any time afterwards be renewed, and again iffued in charge.

Fines uncol-
lected conti-
nued in the
great roll, and
iffued yearly
in procefs,

C H A P. XV.

Of PROFITS of the HANAPER, POST FINES, CUSTODIAM RENTS, FIRST FRUITS, PROFITS upon FACULTIES, and TWENTIETH PARTS.

PRofits of the Hanaper *. This is a duty arifing to the Crown for fealing patents, and for original writs, viz. for all patents or grants of lands or offices, £1 8s. 3d. †,

Profits of the
hanaper what.

* So called from the hamper or bafket, in which original writs relating to the bufinefs of the fubject, and the returns of them, were according to the fimplicity of ancient times kept, as were others, relating to fuch matters wherein the Crown is immediately or mediately concerned, in a little fack or bag, in parva baga; from whence has arifen the diftinction of the hanaper office, and petty bag office, which both belong to the common law court in chancery. 3 Black. 49.

† Of this £1 8s. 3d. the King hath 15 s. (unlefs it be for patents of offices: in which cafe he hath but 1s 6d.) the chancellor 2s. for a docquet; the mafter of the rolls 5 s. the clerk of the hanaper 6s. 3d.

and

and for original writs, differently according to the nature of the writ. But the § moiety of thefe is granted to the Lord Chancellor for the fupport of the dignity of his office.

The clerk of the hanaper paffes his accounts before the commiffioners of the impreft accounts; before whom the books of entries of patents and writs fealed are produced, which are a charge on him for all money received by him; and he likewife fwears to the truth of the charge. His difcharges are the receipt of the Lord Chancellor for his moiety, and his warrants for difburfements for the ufe of the Chancery court, which the Chancellor has a power to make for all or any part of this fund, and Exchequer acquittances for payments into the treafury. And his account is to be figned by the Lord Chancellor yearly.

Post Fines are, as has been mentioned in chap. 13, a duty to the King for a fine acknowledged in the court of Common pleas, to be paid by the cognizee, after the fine is fully paffed; being fo much and half fo much as was paid to the King for the *præ-fine*; and they are eftreated

§ The firft grant of this moiety that I can find was by letters Patents dated 5th of May 1 James I. to Adam Lord Vifcount Loftus of Ely, then Lord Chancellor of Ireland, for the fupport of the dignity of his office, during his continuence therein; and is expreffed to be of one full moiety of all the fines payable to the Crown on original writs,' as alfo of all the profits and emoluments arifing therefrom; which grant was confirmed to the fame chancellor by further letters patents, dated the 5th of May, in the firft year of the reign of King Charles I. but I do not find that this moiety has been expref.ly granted in any of the patents to fubfequent chancellors, altho' it has been conftantly paid to them, unlefs it comes under the general words which have been in all the faid fubfequent grants; to wit, to have, hold, enjoy, poffefs, and exercife the faid office, together with all and fingular the powers, authorities, jurifdictions, immunities, privileges, penfions, fees, falaries, allowances, benefits of the the feal, and iffuable writs, and all other benefits, commodities, emoluments, and advantages whatfoever, to the faid office belonging, incident or in any manner appertaining, or with the faid office, at any time heretofore had, held, or enjoyed.

by

by the court of Common pleas into the Exchequer, and levied by the sheriff of the county off of the lands of which the fine was passed, and answered by him in his account.

CUSTODIAM RENTS. These are such rents as are referved to the Crown on *custodiams*, or leases under the Exchequer seal; which are most commonly made of such lands, *&c.* as are seized into the hands of the Crown upon outlawries in civil actions, whereon, upon motion of the plaintiff in the action to the court of Exchequer, the *custodiam* is given to him towards the satisfaction of his debt, and a small rent is likewise reserved to the Crown.

Custodiam rents on outlawries in civil actions.

And such *custodiams* are also granted by the court of Exchequer for debts due to the Crown; and upon seizures for rents reserved on grants from the Crown of lands, rectories, tithes, *&c.* And in these cases either the solicitor for the King's rents, or the collector of the district, is generally the custodee. For the securing the payment of these *custodiam* rents the custodee gives security by recognizance before the Chief Baron of the Exchequer, who signs the *custodiam*, which is the warrant for its passing under the Exchequer seal.

Or for debts due to the Crown.

These rents are all in charge in the pipe, and from thence process issues to the respective sheriffs for collecting them as he does other sums, on which process the sheriff is apposed at the passing of his accounts, and answers for, and pays them together with the money collected by him on the process of the green wax.

In charge in the pipe.

For more of this matter see chap. CUSTODIAMS.

FIRST FRUITS and TWENTIETH PARTS were alſo branches of the caſual revenue; but they are not ſo at this day. However it may not be amiſs to mention ſhortly from whence they aroſe, and when and how they were diſpoſed of.

Profits upon faculties.

PROFITS UPON FACULTIES. Theſe are ancient profits ariſing to the Crown, being a part or portion of taxes upon the granting of faculties or diſpenſations, according to the allottment thereof by the ſtatute of 28 Hen. VIII. c. 19. revived by ſtat. 2 Eliz. c. 1. to be received and accounted for by the clerk of the faculties; for according to the ſaid act the King is to have ¾ of ⅓ which amounts to £3, and the remaining parts are to be divided among the ſeveral officers in the ſaid act mentioned.

How diſpoſed of.

Of this £3 one moiety was granted by letters patent, dated 10th of April, 20 James I. to Chriſtopher then archbiſhop of Armagh and his ſucceſſors in that ſee; ſo that there remained but £1 10s. to be accounted for to the Crown. And by letters patent, dated 27th of March 1727, 1 Geo. II. one fourth of the money payable for taxes of faculties, rated at and above £4, and which, according to the computation aforeſaid, and in the ſaid act, is the ſum remaining to the Crown, was granted to doctor Marmaduke Coghill, then judge of the prerogative and faculties, in conſideration of his great diligence and trouble in executing the ſaid office, and in regard the ſame was an office of great dignity and conſequence, and required conſtant attendance, and that no ſallary was annexed thereto; to hold to the ſaid Marmaduke Coghill, during ſuch time as he ſhould continue in the ſaid office

of

of judge or commiffary of the courts of prerogative and faculties; under the colour of which grant (for none other appears) the faid fourth part has been ever fince received by the regifters or commiffaries, fucceffors in the faid office, and no part of it accounted for or paid to the Crown.

THE FIRST FRUITS, *primitiæ* or *annates*, are a charge upon admiffion into church livings; being the firft year's profit of every ecclefiaftical benefice or promotion in this kingdom. And they are payable in two years by four gales; for which bonds are taken to the Crown by an officer called the remembrancer, clerk and receiver of the firft fruits; who receives thofe dues and formerly paid them into the treafury; but if default be made in the payment of them to this officer, procefs iffues for the levying them as other bonds to the Crown.

First fruits what.

THE TWENTIETH PARTS were alfo a charge upon all church livings; being the twentieth part of every year's profit of every ecclefiaftical benefice or promotion.

Twentieth parts what.

And thefe profits of the firft fruits and twentieth parts were originally a part of the papal ufurpations over the clergy in this kingdom. And when that power was abolifhed, and the King declared the head of the church, they were annexed to the Crown by the ftat. 26 Hen. VIII. Eng. which is in force here by the 28 Hen. VIII. c. 8. and a valuation was made of them by commiffion grounded on this act, which was entered in what was called the King's book, which was formerly lodged in the chief remembrancer's office in the Exchequer, but now this and all the records belonging to them are in the office of the clerk or remembrancer of them.

Originally papal ufurpations, and transferred to the King as head of the church.

But

But neither of these are at this day any part of the revenue of the Crown; her Majesty Queen Anne having by patent, dated the 7th of February in the tenth year of her reign, released to the bishops and clergy and their successors the said twentieth parts; and having by another patent, of the same date, granted to the several persons therein named and their successors all the first fruits (which are therein said to amount to about £450 a year) in trust for the building and repairing of churches, and purchasing of glebes, where they shall be wanting, and of impropriations, where the benefice shall not suffice; and for the more liberal maintainance of the minister who has the cure of souls. And these patents were by the stat. of 2 Geo. I. c. 15. confirmed; and since the patent and act of parliament, the first fruits are paid by the clerk or receiver of them to the trustees.

CHAP.

C H A P. XVI.

OF WAIFS, ESTRAYS, GOODS of FUGITIVES and
FELONS, DEODANDS, WRECKS, TREASURE-
TROVE, and GOLD and SILVER-MINES.

WAIFS, *bona waviata*, are goods ftolen and waived Waif. or thrown away by a thief in his flight for fear of being apprehended. Thefe are given to the King by law as a punifhment upon the owner for not himfelf purfuing the felon and taking away his goods from him. But waived goods do not belong to the King till feized by fomebody for his ufe; for if the party robbed can feize them firft, though at the diftance of twenty years, the King fhall not have them. And if the goods are hid by the thief, or left any where by him, fo as that he had not them about him when he fled, and therefore did not throw them away in his flight, thofe are not waived goods, and the owner may have them again when he pleafes.

ESTRAYS are fuch valuable animals as are found wander- Eftrays. ing in any manor or lordfhip, and of which no man knows 1 Black. c. 8. the owner; in which cafe the law gives them to the King or his grantee as derelict goods. But in order to veft an abfo-lute property in the King or his grantees they muft be proclaimed in the church and two market towns next ad-joining to the place where they are found; and then if no man claims them after proclamation and a year and a day paffed, they belong to the King or his grantee without re-demption. If the owner claims them within the year and day, he muft pay the charges of finding, keeping, and pro-claiming them.

Waif.
1 Black. c. 8.
5 Co. 109.

GOODS

Goods of fu-
gitives.
Porter 272.
GOODS of FUGITIVES are the goods of a perſon who is found upon record to have fled for felony, whether he be found guilty of the felony or not, which are forfeited to the King as a puniſhment for his having done what in him lay to ſtop the courſe of publick juſtice. But the jury very ſeldom find the flight; the forfeiture being looked upon, ſince the very great encreaſe of perſonal property of late years, as too great a penalty for an offence to which a man is prompted by a natural love of liberty.

Goods of
felons.
˙ GOODS of FELONS are the goods of perſons convicted of felony or treaſon, or put in the exigent, for which ſee chap. 17.

Not to be
ſeiſed until
conviction.
1 Hale 365.
By the ſtat. of 1 Rich. III. c. 3. Eng. it is enacted that neither ſheriff, &c. or any other perſon ſhall take or ſeize the goods of any perſon arreſted or impriſoned before he be convicted of the felony, (under which term Sir M. Hale is of opinion treaſon is comprehended, but qu.) or before the goods be otherwiſe lawfully forfeited, upon pain of forfeiting the double value of the goods ſo taken.

Except where
a ſecond
capias is
awarded.
1 Hale 365.
But by the ſtat. of 25 Edw. III. c. 14. Eng. which is not repealed by the laſt mentioned act, where a perſon is in-dicted of felony, (under which Sir M. Hale ſays treaſon is comprehended) in the ſecond *capias* there ſhall be com-priſed a precept to the ſheriff to ſeize his goods and keep them till the day of return of the writ; and if he be not found then the exigent is to be awarded and the goods forfeited.

James

James late duke of Ormond being attainted of high treafon by an act of parliament in Great Britain, a queftion arofe in the Exchequer here, whether a writ of feizure could regularly iffue to feize his perfonal eftate, until an inquifition was taken and returned. And Lord Chief Baron Gilbert declared that when a writ of feizure iffues it is only to remind the fheriff of his duty; for from the inftant of a perfon's attainder his goods are forfeited to the Crown, and the fheriff *virtute officii* may feize them; and that by the Duke's attainder in Great Britain he was attainted through all the King's dominions; and therefore a writ of feizure was awarded with a claufe of inquiry. 29th of February, 1715.

Writ of feizure awarded upon a parliamentary attainder in Great Britain.

A perfon having been outlawed for treafon, a fpecial writ of *capias ut lagatum* iffued out of the King's bench againft him; and the fheriffs having thereupon returned that he was poffeffed of a cutter lying at one of the quays in Dublin, which they feized for his Majefty's ufe; an application was made to the court for a *venditioni exponas* for the fale of the cutter; but the court held that they could not grant fuch writ, nor do any act for difpofing of the veffel, it having become part of his Majefty's property, over which the court of Exchequer only had jurifdiction. It thereupon became a queftion in what manner the proceedings fhould be removed into the Exchequer; whether by eftreat, or by *certiorari* from the court of Chancery to remove the proceedings into that court, thence to be fent by *mittimus* into the Exchequer. And the court held that it might be either way; but that as the former method was the more expeditious and lefs expenfive, it was therefore the more eligible. And the outlawry was accordingly eftreated. The King againft Connor in the K. B. 1771.

Court of K. B. cannot grant a writ of venditioni exponas to fell goods of an outlaw.

And

And afterwards upon motion in the Exchequer upon the faid eftreat, by the folicitor of the cafual revenue, a writ of *venditioni exponas* was awarded.

Deodands.
1 Black. c. 8.
1 Hale 419.

DEODANDS. By thefe are meant any moveable goods which are the immediate occafion of the death of any human creature, which are forfeited to the King to be applied to pious ufes and diftributed in alms by his high almoner; though formerly deftined to more fuperftitious purpofes. They feem to have been originally defigned in the blind days of popery as an expiation for the fouls of fuch as were fnatched away by fudden death; and for that purpofe ought properly to have been given to holy church, in the fame manner as the apparel of a ftranger who was found dead was applied to purchafe maffes for the good of his foul.

Not forfeited
till the death
be found.
1 Hale 419.

But they are not forfeited till the death be found, which is regularly by the coroner; but may before the commiffioners of gaol-delivery, oyer and terminer, or of the peace, if omitted by the coroner. And the inquifition ought to inquire of the goods that occafioned the death, and the value of them; and the *villata* where the mifchance happened fhall be charged with procefs for the goods or their value, though they were not delivered to them,

Cannot be
inquired of
by the grand
jury at an
affize fecretly.
4 Burr. 19.

But where a man was killed by a fall from a horfe, and the coroner having not taken any inquifition upon the death, the lord of the manor finding himfelf likely to lofe his deodand made his application at the affizes, where the jury found an inquifition or prefentment of the fact; the court of King's bench in Weftminfter-hall quafhed the prefentment, as being a prefentment of entitling tranfacted

in

in fecret, and which the grand jury had no authority to make, at leaft under their general charge from the judge.

And as this forfeiture feems to have been originally founded rather in the fuperftition of an age of extreme ignorance than in the principles of found reafon and true policy, it hath not of late years met with great countenance in Weftminfter-hall; and when juries have taken upon them to ufe a judgment of difcretion, not ftrictly within their province, for reducing the quantum of the forfeiture, the court of King's bench has refufed to interpofe in favour of the Crown or lord of the franchife.

A forfeiture not favoured in courts of law. Fofl. 266. 4 Burr. 17.

WRECK. This, by the ancient common law, was where any fhip was loft at fea and the goods or cargo were thrown upon the land; in which cafe thefe goods fo wrecked were adjudged to belong to the King; for it was held that by the lofs of the fhip all property was gone out of the original owner. But this was undoubtedly adding forrow to forrow, and was confonant neither to reafon nor humanity. Wherefore it was firft ordained by King Henry I. that if any perfon efcaped alive out of the fhip it fhould be no wreck. And afterwards King Henry II. by his charter declared that if either on the coafts of England, Poictou, Oleron, or Gafcony, any fhip fhould be diftreffed, and either man or beaft fhould efcape or be found therein alive, the goods fhould remain to the owners, if they claimed them within three months; but otherwife fhould be efteemed a wreck, and fhould belong to the King or other lord of the franchife. This was again confirmed with improvements by King Richard I. who in the fecond year of his reign not only eftablifhed thefe conceffions by ordaining that the owner, if he was fhipwrecked and efcaped, "omnes res fuas liberas et quietas haberet;" but

Wrecks, the progrefs of the law with regard to them. 1 Black. c. 8.

alfo, that if he perifhed, his children, or in default of them his brethern and fifters, fhould retain the property; and that, in default of brother and fifter, the goods fhould remain to the King. And the laws fo long after as the reign of Henry III. feems ftill to have been guided by the fame equitable provifions; for then if a dog (for inftance) efcaped, by which the owner might be difco-vered, or if any certain mark were fet on the goods, by which they might be known again, it was held to be no wreck. And this is certainly moft agreeable to reafon; the rational claim of the King being only founded upon this, that the true owner cannot be afcertained.

The prefent legal notion of them.

But afterwards, in the ftatute of Weftminifter the firft, the law is laid down more agreeable to the charter of King Henry Ii. and upon that ftatute hath ftood the legal doctrine of wrecks to the prefent time. It enacts that if any living thing efcape, a man, a cat, or a dog; (which, as in Bracton, are only put for examples) in this cafe, and as it feems in this cafe only, it is clearly not a legal wreck; but the fheriff of the county is bound to keep the goods a year and a day, that if any man can prove a property in them, either in his own right or by right of reprefentation, they fhall be reftored to him without delay; but if no fuch property be proved within that time, they then fhall be the King's. If the goods are of a perifhable nature, the fheriff may fell them, and the money fhall be liable in their ftead.

Often granted away by the Crown.

This revenue of wrecks is frequently granted out to lords of manors as a royal franchife; and if any one be thus entitled to wrecks in his own land, and the King's goods are wrecked thereon, the King may claim them at any time, even after the year and day.

It

It is to be obferved that, in order to conftitute a legal wreck, the goods muft come to land. If they continue at fea, the law diflinguifhes them by the barbarous appellations of *jetfam*, *flotfam*, and *ligan*. *Jetfam* is where goods are caft into the fea and there fink and remain under water. *Flotfam* is where they continue fwimming on the furface of the waves. *Ligan* is where they are funk in the fea but tied to a cork or buoy in order to be found again. Thefe are alfo the King's if no owner appears to claim them; but if any owner appears, he is entitled to recover the poffeffion. For even if they be caft over board without any mark or buoy, in order to lighten the fhip, the owner is not by this act of neceffity conftrued to have renounced his property; much lefs can things *Ligan* be fuppofed to be abandoned, fince the owner has done all in his power to affert and retain his property. Thefe three are of admiralty jurisdiction, and accounted fo far a diftinct thing from the former, that by the King's grant to a man of wrecks, things *jetfam flotfam and ligan* will not pafs.

Wrecks, in their legal acceptation, are at prefent not very frequent; it rarely happening that every living creature on board perifhes; and if any fhould furvive, it is a very great chance, fince the improvement of commerce, navigation, and correfpondence, but the owner will be able to affert his property within the year and a day limited by law.

And in order to preferve this property entire for him, and if poffible to prevent wrecks at all, our laws have made many very humane regulations; in fpirit quite oppofite to thofe favage laws which formerly prevailed

U 2 in

in all the northern regions of Europe, permitting the
inhabitants to feize on whatever they could get as lawful prize. For by the ftatute of 2 Ed. III. c. 13. if
any fhip be loft on the fhore, and the goods come to
land, (fo as it be not legal wreck) they fhall be pre-
fently delivered to the merchants, they paying only a
reafonable reward to thofe that faved and preferved them,
which is called falvage. And by the 4 Geo. I. c. 4. and
17 Geo. II. c. 11. further falutary regulations are made for
the encouragement of the affiftance and falvage of fhips
ftranded or in diftrefs.

Treafure
trove.
1 Black. c. 8.

TREASURE TROVE is money, or coin, gold, filver, plate,
or bullion, found hidden in the earth, or other private place,
the owner thereof being unknown; in which cafe fuch
treafure belongs to the King; but if he that hid it be known,
or afterwards found out, the owner and not the King is
entitled to it. Alfo if it be found upon the earth, or in
the fea, it doth not belong to the King, but to the finder
if no owner appears; fo that it feems it is the hiding, not
the abandoning of it, that gives the King a property;
and this diftinction clearly appears from the different in-
tentions which the law implies in the owner. A man
who hides his treafure in a fecret place evidently does not
mean to relinquifh his property; but referves a right of
claiming it again when he fees occafion; and if he dies
and the fecret dies with him, the law gives it to the King
as part of his royal revenue. But a man who fcatters
his treafure upon the publick furface of the earth, or into
the fea, is conftrued to have abfolutely abandoned his
property, and returned it into the common ftock, without any intention of reclaiming it; and therefore it belongs, as in a ftate of nature, to the firft occupant or finder;
unlefs the owner appears and afferts his right, which then
proves

proves that the lofs was by accident, and not with an intent to renounce his property.

GOLD and SILVER MINES are another branch of the royal revenue, which has its original from the King's perogative of coinage, in order to fupply him with materials. By the old common law, if gold or filver be found in mines of bafe metal, according to the opinion of fome, the whole was a royal mine, and belonged to the King. But now by the ftatute of 4 Anne c. 12. no mines of copper, tin, iron, or lead, fhall be adjudged to be a royal mine, altho gold or filver may be extracted thereout. And all perfons that fhall be proprietors of any mines wherein any ore fhall be difcovered, and in which there is copper, tin, iron, or lead, fhall hold and enjoy the fame; but the King is to have the ore at certain prices in the act ftated.

Gold and filver mines.
1 Black. c. 8.

C H A P.   XVII.

Of ESCHEATS and FORFEITURES.

Efcheats what.
2 Blackf.
c. 15.

ESCHEAT is one of the fruits and confequences of feodal tenure, being the determination of the tenure, or diffolution of the mutual ties between the lord and tenant, from the extinction of the blood of the latter, by either natural or civil means. If he die without heirs of his blood, or if his blood be corrupted and ftained by commiffion of treafon or felony, whereby every inheritable quality is entirely blotted out and abolifhed; in fuch cafes the land efcheats or falls back to the lord of the fee, that is the tenure is determined by breach of the original condition expreffed or implied in the feodal donation. In the one cafe there are no heirs fubfifting of the blood of the firft feodatory or purchafer, to which heirs alone the grant of the feud extended; in the other, the tenant by perpetrating an atrocious crime fhows that he is no longer to be trufted as a vaffal, having forgotten his duty as a fubject; and therefore forfeited his feud, which he held under the implied condition that he fhould not be a traitor or a felon; the confequence of which in both cafes is, that the gift being determined refults back to the lord who gave it.

Of two forts, ibid.

ESCHEATS are frequently divided into thefe *propter defectum fanguinis*, and thofe *propter delictum tenentis*; the one fort, if the tenant dies without heirs; the other, if his blood be attainted. But both thefe fpecies may well be compre-

comprehended under the firſt denomination only, for he that is attainted ſuffers an extinction of his blood as well as he that dies without relations. The inheritable quality is expunged in one inſtance and expires in the other; or as the doctrine of eſcheats is very fully expreſſed in Fleta " *Dominus feodi loco hæredis habetur, quɔ̃ es* " *per dɛfectum vɛl delictum extinguitur ſanguis tenentis.*"

EscHE.ATS ariſing merely upon deficiency of the blood, whereby the defcent is impeded, are firſt, when the tenant dies without any relations on the part of any of his anceſtors; ſecondly, when he dies without any relations on the part of thoſe anceſtors from whom his eſtate deſcended; thirdly when he dies without any relations of the whole blood. In two of theſe cafes, the blood of the firſt purchaſer is certainly, in the other it is probably, at an end; and therefore in all of them the law directs that the land ſhall eſcheat to the lord of the fee. For the lord would be manifeſtly prejudiced, if, contrary to the inherent condition tacitly annexed to all feuds, any perſon ſhould be ſuffered to ſucceed to lands, who is not of the blood of the firſt feudatory, to whom for his perſonal merit the eſtate is ſuppofed to have been granted.

Thɛſe defici-ency of blood, ibid.

By attainder for treaſon or felony the blood of the perſon attaintɛd is ſo corrupted as to be rendered no longer inheritable.

Or by cor-ruption of blood, ibid.

But this ſpecies of eſcheat muſt be diſtinguiſhed from forfeiture of lands to the King; which, by reaſon of their ſimilitude in ſome circumſtances, and becauſe the Crown is very frequently the immediate lord of the fee, and therefore entitled to both, have been often confounded

This to be diſtinguiſhed from forfei-ture, ibid.

ed

ed together. Forfeiture of lands and of whatever elfe the offender poffeffed was the doctrine of the old Saxon law, as a part of the punifhment for the offence, and does not at all relate to the feodal fyftem, nor is the confequence of any figniory or lordfhip paramount; but, being a prerogative vefted in the Crown, was neither fuperfeded nor diminifhed by the introduction of the Norman tenures; a fruit and confequence of which efcheat muft undoubtedly be reckoned. Efcheat therefore operates in fubordination to this more ancient and fuperior law of forfeiture.

How it is occafioned, ibid.

The doctrine of efcheat upon attainder, taken fingly, is this; that the blood of the tenant by the commiffion of any felony (under which denomination all treafons were formerly comprized) is corrupted and ftained, and the original donation of the feud is thereby determined; it being always granted to the vaffal on the implied condition of *dum bene fe gefferit*; upon the thorough demonftration of which guilt by legal attainder, the feodal covenant and mutual bond of fealty are held to be broken; the eftate inftantly falls back from the offender to the lord of the fee; and the inheritable quality of the blood is extinguifhed and blotted out for ever.

And operates as to eftates vefted, ibid.

In this cafe the law of feodal efcheats was brought to England at the conqueft, and in general fuperadded to the ancient law and forfeiture; in confequence of which corruption and extinction of hereditary blood the land of all felons would immediately revert in the lord, but that the fuperior law of forfeiture intervenes, and intercepts it in its paffage; in cafe of treafon for ever; in cafe of other felony, for only a year and a day; after which

time

time it goes to the lord in a regular courfe of efcheat, as it would have done to the heir of the felon, in cafe the feodal tenures had never been introduced. And that this is the true operation and genuine hiftory of efcheats will moft evidently appear from this incident to gavel-kind lands, which feem to be the old faxon tenure, that they are in no cafe fubject to efcheat for felony, tho' they are liable to forfeiture for treafon.

Hitherto we have only fpoken of eftates vefted in the offender at the time of his offence or attainder. And here the law of forfeiture ftops; but the law of efcheat purfues the matter ftill farther, for, the blood of the tenant being utterly corrupted and extinguifhed, it follows, not only, that all he now has fhould efcheat from him, but alfo that he fhould be incapable of inheriting any thing for the future. *And as to incapacity of inheriting. ibid.*

This may further illuftrate the diftinction between forfeiture and efcheat. If therefore a father be feized in fee, and the fon commits treafon, and is attainted, and then the father dies; here the land fhall efcheat to the lord; becaufe the fon, by the corruption of his blood, is incapable to be heir, and there can be no other heir during his life; but nothing fhall be forfeited to the King; for the fon never had any intereft in the land to forfeit. In this cafe the efcheat operates, and not the forfeiture; but in the following inftance the forfeiture works, and not the efcheat. As where a new felony is created by act of parliament, and it is provided that it fhall not extend to corruption of blood; here the lands of the felon fhall not efcheat to the lord, but yet the profits of them fhall be forfeited to the King fo long as the offender lives. *Diftinction between forfeiture and efcheat illuftrated, ibid.*

Vol. I.   X   There

Corruption
of blood ob-
structs def-
cent, ibid.
There is yet a further confequence of the corruption
and extinction of hereditary blood which is this; that
the perfon attainted fhall not only be incapable himfelf
of inheriting, or tranfmitting his own property by heirfhip,
but fhall alfo obftruct the defcent of lands or tenements to
his pofterity, in all cafes where they are obliged to derive their title through him from any remoter anceftor.
The channel, which conveyed the hereditary blood from
his anceftors to him, is not only exhaufted for the prefent,
but totally dammed up and rendered impervious for the
future. This is a refinement upon the ancient law of
feuds, which allowed that the grandfon might be heir to
his grandfather, tho' the fon in the intermediate genera-
tion were guilty of felony. But, by the law of England,
a man's blood is fo univerfally corrupted by attainder,
that his fons can neither inherit to him nor to any other
anceftor, at leaft on the part of their attainted father.

confidered as
unreafonable
and unjuft,
ibid.
This corruption of blood thus arifing from feodal
principles, but perhaps extended farther than even thofe
principles will warrant, has been long looked upon as a
peculiar hardfhip; becaufe the oppreffive parts of the
feodal tenures being now in general abolifhed, it feems
unreafonable to referve one of their moft inequitable
confequences; namely that the children fhould not only
be reduced to prefent poverty (which however fevere is
fufficiently juftified upon reafons of publick policy) but
alfo be laid under future difficulties of inheritance, on
account of the guilt of their anceftors. And therefore,
in moft (if not all) of the new felonies created by parliament fince the reign of Henry the VIII. it is declared
that they fhall not extend to any corruption of blood.
But as in fome of the acts for creating felonies, (and
thofe

thofe not of the moft atrocious kind) this faving was neglected or forgotten to be made, it feems to be highly reafonable and expedient to antiquate the whole of this doctrine by one general law.

The natural juftice of FORFEITURE, or confifcation of property for treafon, is founded on this confideration, that he who hath thus violated the fundamental principles of government, and broken his part of the original contract between King and people, hath abandoned his connexions with fociety, and hath no longer any right to thofe advantages which before belonged to him purely as a member of the community; among which focial advantages the right of transferring or tranfmitting property to others is one of the chief.

Forfeiture on what founded
4 Black. c. 29. 1 Black. c. 8.

FORFEITURE is twofold, of real and of perfonal eftates. Firft, as to real eftates; by attainder in high treafon a man forfeits to the King at common law all his lands and tenements in fee fimple; and all leafes for lives or freeholds defcendable; (and all rights of entry thereon) which he held at the time of the offence committed, or at any time afterwards. This forfeiture relates backwards to the time of the treafon committed, fo as to avoid all intermediate fales and encumbrances. But it does not take effect unlefs an attainder be had, of which it is one of the fruits; and therefore if a traitor dies before judgment pronounced, or is killed in open rebellion, or is hanged by martial law, it works no forfeiture of his lands; for he never was attainted of treafon.

of real eftates, by common law, 4 Black. c. 29. Foft. 223.

X 2   And

In what cafes
velled in the
King without
office,
2 Hawk. 448. And the lands fo forfeited by attainder are actually vefted in the King without any office; becaufe they cannot defcend, the blood being corrupted; and they cannot be in abeyance. But by the common law fuch lands were not vefted in the actual poffeffion of the King during the life of fuch offender without an office.

By ftatute. By the 28 Hen. VIII. c. 7. every offender convict of high treafon by prefentment, confeffion, verdict, or pro- cefs of outlawry, fhall forfeit to the King all lands which fuch offender fhall have of any eftate of inheritance 1 Hale. 240. (under which words eftates in tail are comprehended) at the time of fuch treafon committed or after; faving to every perfon other than the offenders, their heirs, and fucceffors, all rights, &c.

And by 27 Eliz. c. 1. all offenders convict of any high treafon, by any act of parliament, confeffion, ver- dict, or procefs of outlawry, fhall forfeit as well all fuch rights, entries, and conditions, as alfo all fuch lands, te- nements and hereditaments, which any fuch offenders fhall have of any eftate of inheritance, in ufe or poffeffi- on, by any right, title, or means, at the time of any fuch treafon committed, or at any time after. And the King fhall be adjudged in actual and real poffeffion of all fuch lands, tenements, &c. of the offenders fo attainted without any office or inquifition to be found of the fame; faving to every perfon other than the offenders in any treafons, their heirs and fucceffors, and fuch perfons as claim to any of their ufes, all fuch rights as they fhall have at the day of the committing fuch treafons, or at any time afore, as if this act had never been made.

Secondly,

Secondly, as to perfonal eftates. The forfeiture of goods and chattels accrues in every one of the higher kinds of offence; in high treafon or mifprifion thereof, petit treafon, felonies of all forts, whether clergyable or not, felf murder or felony *de fe*, and in petty larceny.

Forfeiture of perfonal eftates.
4 Black. c. 29.

There is a remarkable difference or two between the forfeiture of lands and of goods and chattels.

Firft, lands are forfeited upon attainder, and not before; goods and chattels are forfeited by conviction; becaufe in many of the cafes where goods are forfeited there never is any attainder, which happens only where judgment of death or outlawry is given; therefore in thofe cafes the forfeiture muft be upon conviction, or not at all; and being neceffarily upon conviction in thofe, it is fo ordered in all other cafes; for the law loves uniformity.

Lands for-feited only upon attain-der, goods by conviction.

Secondly, in outlawries for treafon or felony lands are forfeited only by the judgment; but the goods and chattels are forfeited by a man's being firft put in the exigent, without ftaying till he is *quinto exactus*, or finally out-lawed; for the fecreting himfelf fo long from juftice is con-ftrued a flight in law.

In outlawries lands forfeit-ed only by the judgment, but goods by the exigent.

Thirdly, the forfeiture of lands has relation to the time of the fact committed, fo as to avoid all fubfequent fales and encumbrances; but the forfeiture of goods and chattels has no relation backwards, fo that thofe only which a man has at the time of conviction fhall be forfeited. Therefore a traitor or felon may *bona fide* fell any of his chattels real or perfonal, for the fuftenance of himfelf and family, between the fact and conviction; for perfonal
property

Forfeiture of lands relates to the time of the fact, of goods to the conviction.

property is of fo fluctuating a nature that it paffes
through many hands in a fhort time, and no buyer could
be fafe if he were liable to return the goods which he had
fairly bought, in cafe any of the prior vendors had committed a treafon or felony. Yet if they be collufively and
not *bona fide* parted with, merely to defraud the Crown, the
law will reach them; for they are all the while truly and
fubftantially the goods of the offender; and as he, if acquitted, might recover them himfelf, as not parted with
for a good confideration; fo, in cafe he happens to be
convicted, the law will recover them for the King.

Of forfeitures on outlawries in civil cafes.
3 Bac. abr. 734.

Befides thofe forfeitures in criminal cafes, there is a forfeiture upon outlawries in civil cafes; for the retiring from
the inquiry of juftice is held fo criminal in the eye of the
law that it is punifhed with the lofs of the offender's
goods and chattels, and the iffues and profits of his real
eftate.

The King thereby ac-quires only a pernancy of the profits. ibid.

But by fuch outlawry the King has no eftate, but only
a pernancy of the profits, nor can he manure or fow the
ground; and his intereft continues no longer than the
party hath an eftate, and determines with the party's
death; and being originally introduced to compel the
defendant to come in the fooner and anfwer the plaintiff's
demand, it may more eafily be fuperfeded or reverfed, and
thereby the King's pernancy of the profits difcharged,
than an outlawry in a capital cafe.

Cattle of a ftranger may be taken on a levari for the King.
Salk. 395.

And the cattle of a ftranger, *levant* and *couchant* on lands
extended on an outlawry, may be taken for the King upon
a *levari facias*, as the iffues and profits of the lands; for
otherwife there might be no iffues at all, or the perfon
outlawed

outlawed might defraud the King of the whole by letting the land to pasturage.

By the bare outlawry the party immediately forfeits his perfonal goods, and they are vefted in the King; but he does not forfeit the profits of his lands nor his chattels - real till inquifition taken. And therefore an alienation *bona fide,* after outlawry and before inquifition, is good to bar the King of the pendancy; but if the outlaw make a feoffment after inquifition, the feoffee has the eftate, and the King fhall have the profits.

Goods forfeited by the outlawry, but profits of lands or chattels real not till inquifition. Salk. 395. Hard. 101. 176. 1 Lev. 33.

C H A P.   XVIII.

Of LANDS PURCHASED by ALIENS.

AN alien born may purchafe lands or other eftates; but not for his own ufe, for the King is thereupon entitled to them. If an alien could acquire a permanent property in lands, he muft owe an allegiance equally permanent with that property to the King of Great Britain, which would probably be inconfiftent with that which he owes his own natural fovereign; befides that thereby the nation might in time be fubject to foreign influence, and feel many other inconveniencies.

An alien may purchafe for the benefit of the Crown. 1 Black. c. 10.

But there muft be an office or inquifition found, to entitle the King to fuch purchafe; for fince the frechold is in the alien, and he is tenant to the lord of whom the lands are holden, it cannot be divefted out of him but by fome notorious act, by which it may appear that the frechold is in another. But if an alien who purchafes lands die, then the

In what cafe there mult be an office found. Co. Lit. 2.

the freehold is in the King, without office found; becaufe no man can take it as heir to the alien, and therefore the freehold is caft upon the King. But if an alien purchafe, and afterwards is made a denizen and then has iffue and dies, the iffue fhall inherit till office found; becaufe there is a perfon in being to take as heir to the denizen, upon whom the law cafts the freehold, which is not to be divefted out of him without the folemnity of an office.

An alien merchant may take an houfe for his abode. Co. Lit. 2. 6. Poph. 36.

An alien cannot purchafe a leafe for years of lands; but if he be a merchant he may take a leafe of an houfe for his habitation for years only; and this is for the encouragement of commerce; but if he depart the kingdom or die, it goes to the King, and not to his executors or adminiftrators; becaufe it was only a perfonal privilege annexed to the alien as a merchant, and which confequently muft expire with him.

The Crown has a right to a difcovery in a court of equity.

The King has the fame right to the aid of a court of equity for a difcovery of the facts on which his title is grounded as the fubject has in ordinary cafes, and founded on the fame principle of juftice, viz. that it is againft confcience for one to enjoy another's property by concealing his property. So determined in the court of Exchequer in Weftminfter-hall, in the cafe of the Attorney general, againft Rofe Dupleffis, Michaelmas 1751, upon an information in the nature of an Englifh bill for an eftate devifed to the defendant who was an alien, and afterwards confirmed upon an appeal to the houfe of lords.

CHAP.

C H A P. XIX.

Of FORFEITURES in MORTMAIN.

THE Clergy in former days had fo great an afcendant over the people by inftilling into them notions of purgatory, and had fo wrought on them by their art and management, that they prevailed on'them to be very liberal of their poffeffions, and efpecially at their deaths to difpofe of them to thofe only who could promife them happinefs in another world. This proving very preju- dicial to the lords, who thereby loft the advantages of wardfhips, marriage, relief, efcheat, &c. (lands in the hands of a religious houfe or perfon being confidered as in a dead hand, *manus mortua*, yielding no fruits to the lord) occafioned the claufe of the ftatute of *magna charta* 9 Hen. III. c. 56. by which it is enacted, that it fhall not be law- ful for any one to give his lands to any religious houfe and to take the fame lands again to hold of the fame houfe, &c.

Occafion of the claufe in the ftatute of magna charta to prevent giving lands to religious houfes, &c.

But aggregate ecclefiaftick bodies found means to avoid this ftatute by purchafing lands holden of themfelves, and by taking long leafes. Alfo all ecclefiaftical fole corpora- tions, as bifhops, &c. thought themfelves out of this ftatute. To meet therefore with thefe evafions the 7 Ed. I. Eng. called the ftatute of mortmain, was made. By which it is provided, that no perfon, religious or other, fhould buy or fell or receive, under colour of a gift or term of years, or

How evaded, but after- wards farther enforced by 7 Ed. I.

any other title whatfoever, nor fhould by any art or inge-
nuity appropriate to himfelf, any lands or tenements in
mortmain, upon pain that the immediate lord of the fee,
or on his default for one year the lords paramount, and
in default of all of them the King, might enter thereon.

Further arti-
fice of the
clergy to
elude the
aforefaid
ftatutes by
feigned reco-
veries.

The clergy when they found themfelves prohibited by
magna charta from purchafing lands, and perceived that
their evafion of that law was provided againft by 7. Ed. I.
began to apply the judgments of the courts to their own
advantage againft the intention of the law; for they
brought their *præcipe* againft the tenant who had agreed
either to give or fell them the lands on demand, and profecuted the fuit as if it had been really an adverfary one;
till the tenant according to the precedent agreement
made default, which was always looked upon as fufficient
ground for a judgment in favour of the defendant. And
the judges, prefuming all recoveries juft and lawful which
were profecuted in the ufual courfe of law, would not
bring thofe covinous ones within the ftatute, though they
were apparently in *fraudem legis*, and attended with all
thofe inconveniencies which thofe ftatutes were made to
prevent.

reftrained by
13 Ed. I.

But the clergy were quickly ftopped in this courfe; for 32
Weftm. 2. 13 Ed. I. Eng. made thefe recoveries by default
to be mortmain; and the expofition of this ftatute by the
judges has been carried as far beyond the letter as their
expofition on 7 Ed. I. feems to have fallen fhort of the
meaning and intention of that law; for though the letter
of this act extends only to recoveries by default, yet they,
and with good reafon, have extended it to all other recoveries, whether by demurrer or verdict or otherwife;
for if thefe fhould not be within the meaning of the act
an

an iffue might be taken fo much in favour of the clergy,
and the evidence offered might be fo weak, that the
whole intention of the ftatute would be eluded.   2 Inft.
75, 429.

Afterwards they found out the method of conveying to
ufes, which was firft introduced to evade the ftatutes of
mortmain and ferved them effectually; for they generally fitting in Chancery, where ufes were folely cognizable, obliged the feoffee to execute the ufe according
to the truft and confidence repofed in him.

Further eva-
fion contrived
by the clergy
by conveying
to ufes.

But this mifchief was provided againft by the ftat. 15
Rich. II. cap. 5. Eng. by which it is declared that if any be
feized of any lands or other poffeffions to the ufe of
any fpiritual perfon, with purpofe to amortize them, and
whereof fuch fpiritual perfon takes the profits, they fhall
caufe them to be amortized by the licenfe of the King
and other lords, or difpofe of them to fome other ufe;
otherwife they fhall be forfeit, according to the form of
the ftatute of 15 Rich. 2. as lands purchafed by people of
religion ; and that no fuch purchafe to the ufe of fuch
fpiritual perfons fhall be thereafter made upon like pain.
And that the fame law fhall likewife be of lands or other
poffeffions purchafed to the ufe of guilds and fraternities.
And that lands purchafed by corporations, or to their ufe,
fhall be within the compafs of the faid ftatute *de religiofis.*

Remedied by
15 Ric. II. 5.

And whereas the ftatutes had been eluded by purchafing large tracts of land adjoining to churches and confe-
crating them by the names of church-yards, fuch con-
trivance is alfo declared to be within the compafs of the
ftatutes of mortmain.

Feoffments to the ufes of bodies not corporate for fuperflitious purpofes reftrained by 23 Hen. VIII. c. 10. Eng.

But during the times of popery, feoffments and other affurances were frequently made of lands, &c. to the ufe of parifh churches, chapels, fraternities, and other bodies not corporate, for fuperflitious purpofes; which though not ftrictly alienations in mortmain were within the fame mifchief; to prevent which, by the ftat. 23 Hen. VIII. c. 10. Eng. it is enacted that all fuch ufes fhall be void.

Salk. 162.

And tho' there is no ftatute to that effect in force in this kingdom, yet it is holden that the King as head of the church and ftate is intrufted by the common law to fee that nothing be done in maintenance or propagation of a falfe religion, and to direct and appoint all fuch fuperftitious ufes to fuch as are truly charitable.

Whether a devife be a conveyance in mortmain.

A devife to a corporation is not, it fhould feem, within the ftatutes of mortmain fo as to entitle the Crown; for fuch devife is by the ftat. 10. Car. I. Sefs. 2. c. 2. void; and therefore the lands fo devifed fhall defcend to the heir at law *.   See Hob 136.

Whether leafes for years are within the ftatutes of mortmain.

It feems not clearly fettled whether long leafes for years are within the ftatutes of mortmain. Brook in his abridgement fays that a leafe for 400 years is, for that 7 Ed. I. mentions a term amongft other contrivances of eluding the law; but that a leafe for 100 years is not within the ftatute, being an ufual leafe. Br. mort. pl. 39. cites 29 Hen. VIII. And in another place he fays that a rent charge for 80 years is within the ftatute. Ibid. pl. 39 cites 4 Hen. 6, 9.

* See the proceedings on the will of doctor Baldwyn, formerly provoft of Trinity college, in the appendix.

C H A P.

C H A P. XX.

Of the MANNER of passing SHERIFFS ACCOUNTS.

THE sheriff is the King's bailiff of his county, and was anciently the receiver of all the King's revenue arising therein. There were several farms of the county that were under his particular care, that is to say, all those farms that were held of the King as of his county. These were under the survey of the sheriff, and he was charged with them, being obliged to answer them in all wants; and for these he pays in his * *profers*, because they were reckoned part of the profits of his bailiwick. But the receipt of all the ordinary or certain part of the revenue is long since turned into other channels, and he is now accountable to the King only for what is called the casual revenue.

The sheriff the ancient receiver of the King's revenue. Gilb Ex. 144. Madox 643.

Every sheriff before he takes upon him the exercise of his office is to enter into a recognizance of £500 with two sufficient ‡ sureties, conditioned that he shall by himself or his attorney make his *profers*, at the Exchequer, on the morrow of the close of Easter and St. Michael, of the issues and profits of his bailiwick; and at Easter term,

His recognizance.

---

* The *profer* was a pre-payment made by the sheriff out of the issues of his bailiwick. Madox 644. For this the sheriffs now pay 15 sh.

‡ But it does not appear that the security has been sued whilst the sheriff has been sufficient.

before

before the Afcenfion, make a § view of his accounts of
the iffues and profits of his faid bailiwick, and fatisfy at
the receipt of the Exchequer all fuch fums of money as
fhall grow due to his Majefty upon the faid view, before
the end of the faid term; and alfo that he fhall appear
as aforefaid before the Barons, on the morrow of All
Souls, to make a true account of the iffues and profits of
his faid bailiwick, and fatisfy and pay all fuch fums of
money, goods, and other cafualties and things, as he fhall
receive or levy in refpect of his Majefty's revenue, or
cafualties whatfoever, &c.

Times of
iffuing the
feveral pro-
cefs to him. And the manner of the fheriff's accounting at this day
is as follows. There iffue twice in every year, viz. in
Hillary and Trinity vacations, the fummonifter's procefs
and the procefs of the pipe, to the feveral fheriffs of the
kingdom; the former for the levying all fines, amerciaments, poft-fines, forfeited recognizances, and fuch like,
which come by eftreat into the Exchequer; and the latter
for levying fuch fums as were nil'd on the fummonifter's
procefs, and formerly for all the certain revenue of the
Crown, fuch as the Crown rents; as it ftill does for
cuftodiam rents, though thefe are accounted a part of
the cafual revenue.

And what is nil'd on the fecond fummons of the pipe is,
as is before mentioned, fent down in the *fchedula pipæ* into
the office of the fecond remembrancer, who thereon, once
a year, viz. in Trinity vacation, fends out the long or
prerogative writ againft goods, body, lands, heirs, executors and adminiftrators.

§ The view was the entrance or forepart of the fheriff's account, which ftood
de bene effe, whilft he was purifying or liquidating it, by producing his warrants and
vouchers, whereby he was to have an allowance or difcharge of any fums charged
on him. Madox 644.

And

And the feveral fheriffs of the kingdom are prefixed on their accounts as follows.

Days of pre-
fixion to the
feveral
fheriffs.

Michaelmas Term.

County of Dublin,
City of Do.
County of Kildare, } on the morrow of All Souls.
County of Meath,
Town of Drogheda,

County of Carlow,
King's County,
Queen's County, } in eight days of St. Martin.
County of Kilkenny,
City of Do.

County of Weftmeath,
County of Louth,
County of Wexford, } on the morrow of St. Martin.
County of Wicklow,

County of Longford,
County of Waterford, } in fifteen days of St. Martin.
City of Do.
County of Cavan,

Hillary

HILLARY TERM.

County of Tipperary,
County of Rofcommon,
County of Leitrim,
County of Down,   } in eight days of St. Hillary.
County of Monaghan,
County of Armagh,

County of Tyrone,
County of Donegal,
County of Limerick,   } on the morrow of the Puri-
City of Do.   fication.
Town of Carrickfergus.

County of Mayo,
County of Fermanagh,
County of Sligo,
County of Galway,   } in fifteen days of St. Hillary.
Town of Do.
County of Antrim,

County of Cork,
City of Do.   } in eight days of the Purifi-
County of Clare,   cation.

EASTER TERM.

City and county of
    Londonderry,   } in fifteen days of Eafter.
County of Kerry,

This

This prefixion being in the nature of a fummons to the fheriff to come in and make his *profers* and account, if he make default, the prefent courfe is to give him four days further, under a pain, to attend. And formerly the practice was to enter thofe fines from four days to four days until three fines, viz. £10, £20, and £40, were fet on him for his default; and then, if he did not attend, an attachment to the purfuivant iffued againft him. But now no more than one fine is impofed in every iffuable term, which is a great indulgence to fheriffs *.

The proceedings againft them for default.

When a fheriff attends to account, he is to be fworn in court by the chief remembrancer to give a true and juft account of all fuch fums of money as he has levied or lawfully might have levied to his Majefty's ufe. (See Dalt. ch. 123.) When this is done, the treafurer's remembrancer enters a rule of courfe for a day for the fheriff's being appofed in court. But before his appofal he is to prepare for the paffing of his accounts in the following manner, viz.

Matters preparatory to the fher.ff's account ng.

He is firft to make a copy of all the procefs which has iffued to him from the feveral offices in a book for that purpofe.

Making copy of procefs.

Then to go to the fummonifter and to compare with him all the procefs iffued from that office, and to mark the fums he *tots* himfelf with in the margin of his book.

Comparing, &c. with the fummonifter

---

* Anciently it feems in this kingdom (and it is faid the courfe is now fo in England) if the fheriff did not attend his day of prefixion, £5 per day being fet on him as a fine for four days together for his default, then an attachment, and alfo a feizure *nomine diftriƈƚionis* iffued againft him for his non-attendance. See Madox 644. Gilb. Treat. of the Exch. 146.

And with the foreign appofer.

Then he is to bring the fame procefs to the foreign appofer and compare and lodge it with him.

And with the clerk of the pipe.

Then to bring all the procefs of the pipe to the clerk of the pipe, *totting* as aforefaid.

And with the comptroller.

Then to bring the fame procefs to the comptroller of the pipe and compare with him, *totting* as aforefaid, and lodge the procefs with him.

And with the treafurer's remembrancer.

Then to bring all the procefs iffued from the treafurer's remembrancer and compare with him, *totting* as aforefaid, and to lodge the procefs with him, as alfo an inquifition which the fheriff muft take in the county, &c.

And with the clerk of the firft fruits.

Then to go to the firft fruits office, and if he has any procefs from thence he is to examine and lodge them there.

Giving notice of his appofal.

And he is to give notice to each office as he paffes through, as alfo to the Auditor general, and the clerk or Solicitor for the cafual revenue, of the day of his appofal.

Auditor general's duty during the appofal.

The Auditor general is to fit in court during the appofal of the fheriff to take an account of the *tots* in the feveral procefs, and to caft them up, and to give in the total fum to the treafurer's remembrancer, who enters it in his book.

The court will give neceffary orders.

And during the account the court will give fuch orders as are requifite for the fecuring of the faid debts, or for the releafe of the fubject. If the fheriff make an infufficient or an unfatisfactory anfwer they will order him to *tot* for fuch charge. But in fuch cafe the fheriff may have

a writ

a writ of affiftance to levy the money; as he may in all cafes where he charges himfelf with any money he has not received.

After he is appofed, he is to get a *conflat* of his fines and profers from the Chief remembrancer, a *conflat* from the treafurer's remembrancer of the fums charged in his pro- cefs, a *conflat* from the fummonifter and clerk of the eftreats of attainders, if any, if not, that there are none; and a *conflat* from the clerk of the firft fruits; which four *conflats* muft be lodged with the Auditor general, together with his own certificate of what waifs, eftrays, felons or fugitives goods, if any, came to his hands, or if none, a negative certificate. Then he is to get a tranfcript upon thofe *conflats* and certificate from the Auditor general, and a tranfcript alfo from the foreign appofer of what he charged himfelf with in the fummonifter's procefs; which two tranfcripts are to be figned by the three Barons, and the tranfcript from the Auditor is to be entered with the fecond remembrancer. Then he is to get his certificate of * allowance from the fummonifter and clerk of the eftreats, and a *conflat* from the comptroller of the pipe of the fums charged in his procefs; all which are to be fixed together and lodged with the clerk of the pipe, who then makes out the fheriff's † *debet* thereon, which fhows what money he

Conflats to be got by the fheriff after appofal.

Z 2

---

* This is an allowance of 5l. 2s. each feffions for treating the juftices and clerk of the peace at the quarter feffions; provided that eight of the juftices certify at the foot of the eftreat that the fheriff expended fo much. And this allowance is to be out of the feffion fines if fo much be folvent; if not, the deficiency is to be out of his own pocket.

† It is worth obferving that the clerk of the pipe omits inferting in thofe *debets* all fuch fums which the fheriff *tot* for in the fummonifter's procefs, and which he con- ceives to be difchargeable or reducible; which not only is a confiderable lofs to the crown

he is to pay.   Then he is to bring the *debet* to the treafury, pay in the money, and get an Exchequer acquittance
for it, which he is to bring back to the pipe office, to be
annexed to his account.   And if there be no charge ftanding out on his account he may have his *quietus*; but if
there be, he cannot have it until he has fully cleared his
account.·

Old practice,
as to com-
pelling sheriff
to pay his
tots &c by
fine and at-
tachment.

Formerly the fheriff after appofal had but fix days by
the rules and courfe of the court to bring in his accounts
from the feveral offices, to get his debet from the pipe office,
and pay his tots into the receipt of the Exchequer.   And
if he neglected to pay his tots accordingly, the clerk of
the pipe having certified his default to the treafurer's remembrancer, the courfe was to fet three fines upon him,
giving him four days between each fine ; and if he ftill
neglected, then there went an attachment to the purfuivant againft him.

No fheriff to
be attached
for any ne-
glect relative
to his account
but by writ,
or warrant.

And the practice was, when the fheriff was brought
in on fuch attachment, to make him account in cuftody
and not depart till he had finifhed it.   *

And

cafual revenue, but tends to fruftrate the execution of juftice, as forfeited recognizances and fines impofed on officers of juftice for breach of duty, and on other offenders, are contained in this procefs. Befides, feveral of thefe fums have been actually levied, and are in the hands of the fheriffs ; and the delinquents, from the length
of time fheriffs have to pay in their *tots*, have had full opportunity of applying to
reduce or difcharge them.   Wherefore, it feems proper, that the clerk of the pipe
fhould include thefe *tots* in the fummonifter's procefs in the *debet*, as well as others,
or as the fums which are *totted* for in the procefs of the pipe, and of the fecond
remembrancer, if not difcharged or reduced at the time he delivers his *debet* ; as
by not doing fo, the tots in this procefs are attended with the fame mifchief as the
onies in the pipe procefs, *viz* the keeping the accounts of fheriffs ftanding out for
years.   Note, poft fines for licences to accord are alfo in the fummonifter's procefs.

    * Formerly when an attachment iffued to the purfuivant againft a fheriff for not
accounting, not paying his tots, or not clearing his accounts; he could not be difcharged

And by a general rule, fheriffs who neglected to pay Ibidem. their tots, or finifh their accounts in due time were to ftand committed, and the purfuivant was to take them into cuftody, without further order, to prevent which there is a claufe in the ftat. 12 Geo. I. c. 4. that no fheriff or fubfheriff fhall be attached by any officer of the court of Exchequer or other perfon, for not being appofed on any writ or procefs, for not finifhing his accounts in due time, or for any contempt or neglect relating to his account, but by writ under the feal of the court, or by warrant for that purpofe to be figned by the Lord Chief Baron, or in his abfence by either of the other Barons, to be executed by the purfuivant of the court or his deputy, in which warrant the name of fuch fheriff, &c. fhall be particularly inferted and his offence particularly fpecified.

Afterwards * this practice was altered, and when the Altered to feveral fheriffs came upon their accounts, the court gave fines *ad inf-* them four days for paying into the treafury the feveral *nitum.* fums

---

charged therefrom but by a *fuperfedeas*, in confequence of an order of court for the purpofe, on his fhewing, by the treafury acquittance or clerk of the pipes certificate, that the tots were paid, and lodging the fame with the fecond remembrancer ; as appears amongft others, by the rules of this court of the 31 Jan. 1737, 21 and 25 July 1739 and 25 Apr 1740. Whereas of late years the purfuivant has taken upon him to difcharge fuch perfons from attachments, without fuch authority, upon lodging the money with him ; whereby the accounts in fuch cafes have ftood uncleared for years ; and the money all the time remained in the purfuivant's hands.

* It may be here obferved that until the year 1704, it feldom appears that more than three fines were impofed upon fheriffs either for not accounting, not paying their tots, or not finifhing their accounts, but then the fines began to be more excelfive, and the procefs of attachment was neglected ; and from the year 1709,

fums with which they had totted themfelves; which if they failed to do, the court at the inftance of the fecond remembrancer, the firft of the eight days after each iffuable term, entered of courfe a conditional fine of five pounds on them, unlefs they paid their tots in four days; which if they neglected to do, that fine was made abfolute and a further fine impofed, which was always double the laft; and at the end of the term thefe fines iffued in procefs to the fucceeding fheriffs. If the fheriffs did not pay in their tots before the following iffuable term, the court on motion began to fine them *de novo*, and continued fo 'to do every fuch term, until their tots were paid in and their accounts cleared; which fines have fome times amounted to £1200 and upwards, and were, as the former, iffued in procefs to the fucceeding fheriffs.

1709, until lately, fines were fet on fheriffs without end for their neglect, and no other procefs whatfoever iffued to compel them to this part of their duty. This practice was of great prejudice not only to the revenue but likewife to the fheriffs; for after they had been appofed in court they frequently left the profecuting their accounts to their fubfheriffs, who having got the king's money into their hands neither finifhed their accounts nor paid the tots into the treafury. And fines only being fet on them for their neglect, and thefe going in procefs to the fucceeding fheriffs, the fubfheriffs for the time being, to indulge the preceding fubfheriffs, without the knowledge of the fheriff againft whom the fine iffued, om'd on their accounts for thofe fines from time to time; by which means the Crown was not only kept out of them, but fometimes by the death of the fheriffs or their fureties they were entirely loft, and the fheriffs being often deceived by their fubfheriffs, who informed them that every thing had been done, never heard of the fines againft them until perhaps the fubfheriff and his fureties died or became infovent; fo that the fheriff found himfelf loaded with heavy fines, and was left to pay his tots and difcharge his account at his own expenfe; all which would have been prevented, if the regular procefs of the court had iffued againft him.

But

But fince the ftat. of 23 Geo. II. c. 13. the fecond re- Late practice.
membrancer does not iffue any fine againft a fheriff for
not accounting, not paying his tots or not clearing his accounts, until fix months after his appofal, and even then not
but in an iffuable term; fo that if the appofal be in Hillary
term, the firft fine will not be until the Hillary term following; and as an attachment is not to iffue until after a
third fine is eftablifhed, and as the fines are impofed but
every iffuable term, it will be upwards of two years before the attachment iffues. The firft fine impofed is £10,
the fecond £20, and the third £40. But this practice,
which is productive of very great inconvenience and delay,
feems to be founded on a mifconftruction of this act,
which relates to fubfheriffs only.

The fheriffs in their accounts totting themfelves with Proceedings
fome particulars, and in others *onying*, the courfe has againft them
been, to give them to the end of the next iffuable term for fums
after their appofal to procure receipts and other vouchers oni'd.
in difcharge of the fums o*'t* for; and if they do not
procure them within that time, then fuch fines are entered and fuch procefs are fent againft them, as where
they do not account or pay their tots; and for impofing
fuch fines and iffuing fuch procefs againft them for fuch
their neglect, the clerk of the pipe's certificate is the treafurer's remembrancer's warrant *.

By

---

* Here likewife has been of late great neglect in not purfuing the ufual method, and taking the regular procefs for compelling the fheriffs to clear their
accounts; (for proceeding againft them by fines only has been found ineffectual,
as the fheriffs, to whom fuch fines iffued in procefs, oni'd for them, and their fucceffors might oni for them again, and fo on *ad infinitum*;) whereby the Crown is
defrauded,

Poundage al-
lowed the
fheriff on
debts to the
Crown col-
lected by him.

By the ftat. 12 Geo. I. c. 4. all fheriffs who fhall levy any debts, duties, or money, (except poft fines) due to his Majefty, by procefs to them directed upon the fummons of the pipe or green wax [*or*] by *levari* out of the Exchequer, fhall have an allowance on the accounts of 12 d. out of every 20 s. for any fum not exceeding £100 by them levied and collected; and of 6 d. for every 20 s. over and above the firft hundred pounds; and for all debts, duties, and fums of money, (except poft fines) due to his Majefty, by procefs [*ou*] *fieri facias*, and extent iffuing out of any of the offices of the court of Exchequer, one fhilling and fix pence out of every 20 fhillings for any fum not exceeding one hundred pounds by them levied or collected, and 12 d. for every 20 s. over and above the firft £100 * provided fuch fheriff fhall duly anfwer for the fame upon his account by the day on which he ought to be difmiffed the court, or in fuch time to which he fhall have a day granted to finifh his account.

When

defrauded, and the publick greatly injured. For the fums ufually *oni'd* for, being fines and amerciaments fet on fheriffs and other officers by courts of juftice for neglects and mifdemeanors, and recognizances forfeited in the King's bench and at affizes and feffions, which were intended as punifhments, are by means hereof rendered vain and fruitlefs. And fometimes in fact thefe fines *oni'd* for have been received by the fheriffs, and by neglect of calling on them regularly to clear their accounts are funk in their pockets and converted to their own ufe; but this inconvenience has been in a great meafure remedied by a refolution lately made by the court of Exchequer, of not permitting fheriffs to *oni* on their accounts for former fheriffs, but making them either tot for fuch fines, or return inquifitions finding the eftates of the perfons fo fined.

* This claufe, by not being faithfully copied from one in the ftat. 3 Geo. I. c. 15. Brit. of which it was intended to be almoft a tranfcript, is not fenfe; the word *or* being thro' miftake omitted, and the word *and* being fubftituted in

When it appeared to the court that the fubfheriff had
received the King's money on the procefs a..d neglected
to pay it, it has been ufual for the court, at the inftance
and in aid of the fheriff, to grant an attachment againft
the fubfheriff.

Penalties on
fubfheriffs re-
ceiving King's
money and
not account-
ing.

And by the 23 Geo. II. c 13. if any fheriff of any
county, or county of city or town corporate, fhall pay his
fubfheriff or attorney any money, in order to be by them
paid over in difcharge of the accounts of fuch fheriff,
and fuch fubfheriff or attorney fhall neglect to pay over
into the treafury the fums fo to them intrufted, or which
fuch fubfheriff fhall receive on account of fuch fheriff, and
to procure to be taken off, at their own cofts and charges,
all fines laid on fuch fheriff, on account of his not paying in
the fums fo received, within fix months from the time that
any fuch fums fhall be fo paid to fuch fubfheriff or attorney,
fuch fubfheriff, or attorney, fhall for ever after fuch failure

As alfo on
the attornies
of fheriffs.

in place of *on*. Whereby an inftance is created amongft many others that might
be produced, of the inattention too often given to the framing and wording Irifh
ftatutes. In the laft feffion of Parliament, heads of a bill were brought into the
Houfe of Commons and paffed there, *for the improvement of the cafual revenue and
for the better execution of publick juftice*, by giving fheriffs, as a further encouragement to collect thefe branches of it, five fhillings in the pound, for the
fums fo collected; but thefe heads of a bill were thrown out by the Houfe of
Lords. Such an act, it is thought if properly framed, befides improving the cafual
revenue, would be a great means of reftraining the many riotous diforders and
flagrant breaches of the publick peace throughout the kingdom. It would
encourage the fheriffs to collect thefe cafualties, which would in time prevent the
offences, inftead of trufting this moft important matter wholly to their bailiffs,
as is the cafe at prefent, and fwearing a pofitive oath upon their appofal, on the
return made by thefe low people; who, as well fubfheriffs (it is well known)
make largely thereby.

or neglect, be disabled to take or execute the office of subsheriff.

Time may be granted upon motion. But it is thereby provided, that if any subsheriff or attorney shall apply by motion to the court to enlarge his time for paying in such money, and taking off such fines, the court may, upon proof by affidavit that the sheriff to be affected by such motion had due notice of such intended application, examine into the matter, and thereupon grant to the person so applying such further time for paying in the sums, and taking off the fines before mentioned, as to them shall appear reasonable; and in case the sums so received shall not be paid in, and all such fines taken off, within the time so allowed by the court, every subsheriff and attorney shall be liable to the penalties and disabilities aforesaid.

Forfeiture for taking on the office after incuring the disabilities. And if any person who shall incur the disabilities aforesaid, or either of them, shall take upon him the office of subsheriff of any county or city, and be thereof convicted, he shall for every such offence forfeit £500, one moiety to his Majesty, and the other to such person as shall sue for the same.

Procefs to iffue againft Executors, &c of sheriffs dying. Henry Clarke, Esq; late sheriff of the county of Louth having died in office, ordered that *scire facias* do issue against his executors or administrators to compel them to enter on his accounts, 23 Jan. 1710 *.

* So where there are joint sheriffs, as in corporations, and one only has acted, upon an affidavit thereof the acting sheriff only shall be admitted to account; so where one has died who never acted, the survivor upon such affidavit shall also be admitted to account, without any *scire facias* against the executors or administrators of the deceased.

N. Loftus,

N. Loftus, Efq; late fheriff of the county of Wexford Sheriff ad-
having been in a bad ftate of health, during his fheriff-
alty, and ftill continuing fo, and not having intermed-
dled with the green wax procefs, or any other bufinefs of
his office, which his Majefty was entitled to any account
of, upon affidavit thereof, and that he was willing to pay
all his *tefts*, and fuch other demands as his Majefty might
have on account of the faid office, his perfonal attendance was difpenfed with, and his fubfheriff admitted to
pafs his accounts, his Majefty's attorney general confenting thereto; which it feems the court required, altho'
the officer faid that he had not known an inftance before
where, in fuch cafe the confent of the attorney general was
required. 22 June 1765 †.

Upon an application of the late fheriffs of the county Sheriffs not
of the town of Drogheda, to be excufed from appearing excufed from
and accounting, on an affidavit that they had not received the procefs
any of the King's procefs, the court refufed the mo- not being de-
tion; Baron Power obferving that every fheriff is an- them by the
fwerable *fub nomine vice comitis* for, and muft pay in his purfuivant.
proffers, and iffues, whether he receives any of the
procefs or not; and all waifs, eftrays, goods of felons,
and fugitives, &c. not granted away by the Crown, muft
likewife be accounted for by him as fherilf. Every fheriff therefore, as the King's ancient bailiff of his revenue, is bound by law to account, though he happen not
to have received any of the procefs; but the not re-

---

† See Maddox 659, 659. &c. feveral inftances where fheriffs and other officers
accomptants to the King. were admitted to account by their attornies, and fee
the form of the fheriffs recognizance before.

ceiving

ceiving fuch procefs is an excufe for not collecting thofe
fines and other debts which can only be levied under
fuch procefs: But if, on account of the neglect of the
purfuivant in not delivering any of the procefs, to
the fheriff the court fhould excufe the fheriff from
appearing and accounting, not only the certain annual
revenue paid unto the Crown (anciently and at this day
called the fheriffs *proffers*) but the aforefaid cafualties in
feveral inftances would be loft, and many other obvious
ill confequences highly prejudicial to the cafual revenue
would follow.   Mich. 1772.

Sheriff having
obtained his
quietus not to
be called in
queftion after
four years.

By the ftat. 7 W. III. c. 13. every fheriff who fhall
pafs his accounts, and have his *quietus eft*, his heirs, executors, &c. fhall be difcharged of all fums of money
which he fhall have levied or received, and pretended not
to be accounted for, unlefs fuch fheriff fhall be called in
queftion within four years after the time of fuch account
paffed, and *quietus eft*.   And every officer, who fhall fend
out any writ or procefs, or by whofe default any writ or
procefs fhall be fent out, contrary to the act, fhall for
every fuch offence forfeit to the party grieved £40, with
his cofts and damages, to be recovered in any of his
Majefty's courts of record in Dublin.

Perfons ob-
ftructing fhe-
riffs in the
paffing their
accounts to
make fatis-
faction.

And by the 12 Geo. I. c. 4. if any officer or other
perfon concerned in the paffing fheriffs accounts, fhall
wilfully retard or hinder any fheriff in the paffing his accounts, or by his wilful neglect, abfence or other undue
means, prevent any fheriff from being appofed or caft
out of court in due time, or, after payment or tender of
their due fees in faid act afcertained, fhall neglect to enrol, make out, fign, and deliver his *quietus* in due time,
in every fuch cafe the perfons fo offending fhall make
                                                         fuch

fuch recompence to the party aggrieved, as fhall be ordered by the barons, upon complaint exhibited to them in
a fummary method.

Whereas the ufual practice of the court is, that all Rule, 19th
Nov. 1767.
fheriffs of the kingdom fhould pay in or legally difcharge
all fuch fums of money as upon their appofal they *tot*
themfelves with, within fix days of their being fo appofed, and the court taking notice that the feveral fheriffs
do, after their being fo appofed, neglect profecuting, paying, or difcharging their faid *tots*, for confiderable times,
&c. to his Majefty's apparent prejudice, &c. It is ordered
that every fheriff, who, within fix * days after appofing
as aforefaid, does not pay or legally difcharge his *tots* as a-
forefaid, ftand attached by the purfuivant of the court,
until the further order of the court; whereof the faid
purfuivant is not to fail, but to attach the faid fheriff by
virtue hereof, as often as occafion.fhall require as aforefaid.

Whereas by former † rules of the court the fheriffs Rule, 13th
May 1672.
of the feveral counties, cities, and townfhips of this kingdom were required to enter the names of their feveral
attornies, and file their warrants of attorney in the court,
to the end his Majefty's officers may know when to call
for fuch returns from faid fheriffs, as they make on his
Majefty's procefs, which the faid fheriffs hitherto failing
to do. It is now ordered for the advancement and fur-
therance of his Majefty's fervice, that every attorney of

---

* This rule is not now purfued, fee before, p. 172.

† Upon the ftricteft fearch for feveral years before, I cannot find any rule
to this purpofe.

the

the court, who is appointed for any of the said sheriffs, do, within six days after the receipt of their warrant of attorney file the same in this court; and every attorney that shall fail therein is to forfeit unto his Majesty the sum of £5 sterl. as a fine. And it is further ordered that all sheriffs who shall fail in electing and making their attornies, and granting and sending them such warrants as aforesaid, shall be proceeded against according to the statute in that case made and provided.

Rule, 26th May 1684. Upon motion of the attorney for the commissioners of his Majesty's revenue, that several of the sheriffs of this Kingdom, are very slow and remiss in prosecuting the clearing their accounts, or paying in their *tots* after they are appofed, tho' the court never gave them above six days after their appofal to pay in their *tots*, &c. Ordered that if any sheriff, who shall be appofed after this day, shall be remiss, and not pay in his *tots* in six days * after appofal, that then attachments to the pursuivant shall of course issue against him that shall be so in contempt as aforesaid.

Rule, 26th April 169. It is this day ordered by the court, that their Majesty's Auditor general do, for the future, upon the passing any sheriff's accounts, give notice to Mr. Richard Thompfon, and Mr. Chetwood, who are concerned for the commissioners of his Majesty's revenue, upon the passing the sheriffs accounts in the said office, to the end the said sheriffs may be fully charged for the cattle, &c.

* Vide ante.

Upon

Upon motion of Mr. Howard, folicitor for the cafual Rule, 3d revenue, it is ordered for the future, that the fheriffs of June, 1761. this kingdom do give notice in writing to the folicitor for the cafual revenue, of their paffing their accounts, to the end that the faid fheriffs may be fully charged.

Where a fheriff hath *totted* or *oni'd* for a former fheriff, Writ of affiftance in aid of the fheriff when to iffue. or charged himfelf in his accounts, for any perfon what- foever, and hath not levied the fums he fo charged himfelf with, he may apply to the court for a writ of Affiftance, which is granted upon the motion of an attorney and an affidavit at the foot of the fchedule, from the pipe roll, of the feveral fums the fheriff charg- ed himfelf with, that fuch fums nor any part thereof have or hath been received by him, or by any perfon for his ufe, but are ftill ftanding out ; and fuch writ of *affiftance* is in the nature of an execution, to en- able the fheriff to levy the fame, of the body, goods, and lands of the perfons for whom he fo *oni'd*, or charg- ed himfelf, and an enquiry may be held thereon; and if lands be found upon fuch inquifition returned, a cuftodiam fhall be granted.

But thefe *onies* for former fheriffs feldom happen of late, as inquifitions finding their eftates are now expect- ed by the court upon the procefs of the pipe, purfuant to the aforefaid rule of the 24th of February 1695, which rule had a long time been neglected, and thefe *onies* carried on from fheriff to fheriff for a courfe of years, to the great diftrefs and lofs to feveral fheriffs.

And

Inquifition and cuftodiam thereon.

And this inquifition and the cuftodiam thereon are to be proceded upon in the fame manner as all others are; and when the fheriff is fatisfied the fums for which the writ of *affiftance* iffued, either by having received them, by their having been difcharged by the court, or paid into the treafury by the perfons chargeable therewith, the cuftodiam may be diffolved on a confent of the attorney for the fheriff at whofe fuit the writ of *affiftance* had iffued, the rule of which is entered of courfe.

Cuftodiam a-gainft fheriffs for not clearing their accounts.

So, where the cuftodiam is upon procefs at the fuit of the Crown, againft the fheriff for not clearing his accounts, on the fheriffs againft whom the fame had iffued, difcharging the fines and reducing the recognizances for which the *oxies* were, and paying the reduced fums into the treafury, an order is conceived, on confent of the folicitor for the cafual revenue, ·that the cuftodiam fhall be diffolved, which rule is alfo entered of courfe.

Fines impofed on fheriffs, how reduced.

So, where fines have been impofed on fheriffs for non-execution, or for mif-execution of any procefs directed to them from any of the courts above, and that the fame have been eftreated and iffued in procefs, they may upon fufficient caufe fhewn, or by confent of the attornies who iffued the writs, be reduced by the commiffioners of reducement to whatever fums the faid commiffioners may think proper; and then, upon mo·tion of the fheriff's attorney, an order is conceived that the clerk of the pipe do make out a *debet* of the fum to which the fines were reduced, for paying of the fame

fame unto his Majefty's receipt, and on payment thereof that the faid fines be abfolutely difcharged; and if fuch fheriff hath taken any bonds, bills, or any other fecuri- ties for, or on account of, the faid fines, he is to re- ftore the fame; or if he hath the body of the fheriff on whom they were impofed in cuftody for that caufe only, he is on fight of the faid order to enlarge him, and the prefent fheriff is to be thereof exonerated on his ac- counts.——But note, this application to the court of reducements is not neceffary but in the cafes of foreign fines, which (as has been faid before) are fines for the fame caufes, impofed by the other two fuperior courts of record, and eftreated into this; for the fines im- pofed in fuch cafes by this court, where the writ has iffued from it, may be difcharged, on fuch confent as afore- faid, of the plaintiffs attorney, he being one of its own officers, on paying fome fmall fum, as is before menti- oned, into the poor box.

Whereas it appeared, on the appofal of the fheriffs of the counties of Leitrim, and Rofcommon, that the green wax procefs, which iffued out of the proper offi- ces had not been delivered by the purfuivant to them, and the court being informed that by a ftanding rule * thereof, affidavits fhould be made of the delivery of the procefs to the refpective fheriffs; it is therefore this day ordered, that the purfuivant attending this court do on the firft day of every Michaelmas term file an

Rule, 25th Feb. 1774.

* Mr. Howard, who fearched the books for above an hundred years could not find fuch a rule.

† affidavit in the proper office, fetting forth the refpective times when, and to whom, the faid green wax
procefs were delivered by him or his meffenger.

† It is impoffible for the purfuivant to make this affidavit unlefs he delivers
them himfelf, which is almoft impoffible; and that his meffengers fhould make
fuch affidavits, would be attended with no fmall expenfe; and if the purfuivant
does his duty they are unneceffary.

N. B. This rule fhould have been in chap. 14, but was not known when that
chapter was in the prefs.

CHAP.

C H A P. XXI.

Of the FORFEITURES in this KINGDOM by the REBELLION in 1641.

THERE were two confiderable forfeitures in this kingdom, which, as they were attended with very peculiar circumftances, and as feveral acts of parliament were made relative to them, under which a great part of the landed property of this kingdom is derived, deferve a more particular enquiry.

Two confiderable forfeitures in this kingdom.

Of thofe two forfeitures the firft was occafioned by the rebellion which broke out in this kingdom on the 23d of October, 1641 ; the other by the rebellion in 1688, after the abdication of King James and the revolution in favour of William III. and Queen Mary.

At what times.

The former of thefe rebellions was begun and carried on, whilft the civil war was fubfifting in England between King Charles I. and his fubjects; and was afterwards quelled, and the kingdom reftored to peace, during the ufurpation of Oliver Cromwell, and before the reftoration of King Charles II. by the contributions and affiftance of feveral of his Majefty's fubjects both in England and Ireland.

The rebellion in 1641.

And numbers of perfons, who had eftates in land and other properties, having been engaged and concerned in this rebellion, the forfeitures to the Crown were fo confiderable that the interpofition of the Englifh parliament was by it judged neceffary; and accordingly by an act

Act of 17 Car. I. Eng. for the encouragement of adventurers,

B b 2   paffed

paſſed there, 17 Car. I. it was amongſt other things
enacted, that all ſuch rights, titles, intereſts, &c. as
the ſaid rebels, or any of them, on the ſaid 23d of October,
1641, had, or afterwards ſhould have, in any lands or
other hereditaments, ſhould be forfeited to his Majeſty,
and ſhould be deemed, adjudged, veſted, and taken to be
in the actual and real poſſeſſion of the ſaid King, his heirs
and ſucceſſors, without any office or inquiſition thereof to
be found.

And reducing the rebels. And for reducing the rebels, and diſtributing their lands
amongſt ſuch perſons as ſhould advance money and become
adventurers in the reduction, two millions and a half of
acres were to be aſſigned and allotted in this proportion,
viz. each adventurer of £200 was to have 1000 acres in
Ulſter; of £300, 1000 acres in Conaught; of £450,
1000 acres in Munſter; and of £600, 1000 acres in
Leinſter; according to Engliſh meaſure. And the bogs,
woods, loughs, and barren mountains, were to be caſt
into theſe two millions and a half of acres, and ſo thrown
into each man's diviſion.

Quit rents. And out of thoſe acres there was to be a yearly quit-rent
reſerved to the Crown, viz. one penny in Ulſter, three half-
pence in Conaught, two pence farthing in Munſter, and
three-pence in Leinſter.

Survey and allotment directed. And by the ſaid act a commiſſion was to iſſue to ſurvey
all the lands of the rebels that ſhould be forfeited, and to
meaſure 625,000 acres in each province, caſting in bogs,
mountains, &c. as above. And theſe lands were to be
divided amongſt the adventurers by equal lot by the
commiſſioners appointed under the great ſeal; and each
allotment was to be returned into the high Court of
Chancery.

Chancery. And every adventurer, by fuch allotment, was to be in actual feizin of his fhare.

And by the act every perfon within three months after allotment that fhould have 1000 acres in Leinfter, 1500 in Munfter, 2000 in Connaught, or 3000 in Ulfter, was to have power to erect a manor, with a court baron and a court leet, with all other privileges belonging to a manor, and with deodands, fugitives goods, &c.

Adventurers of a certain quantity to have manors, &c.

In the year 1652, the kingdom being reduced and the rebellion ended, the Englifh parliament publifhed an ordinance, called, an ordinance for the fettling of Ireland; in which, declaring that it was not their intention *to extirpate the whole nation*, almoft all the papifts of the kingdom who were worth 10l. were divided into four claffes.

Ordinance, dividing moft of the Irifh into four claffes.

Firft, all perfons who before the 10th of November, 1642, had contrived, acted, or aided the rebellion, murders, or maffacres, which began in October, 1641, and all jefuits, priefts, &c. who had any way contrived, aided, or abetted, and all perfons who fince the 1ft of October, 1641, had flain any perfon not bearing arms for the Englifh, or who, not being then maintained in arms under the command and pay of the Irifh againft the Englifh, had flain any perfon maintained in arms for the Englifh, and all perfons, who being in arms againft the parliament of England, fhould not lay them down in twenty-eight days, and fubmit to their authority, were excepted from pardon of life or eftate.

Firft clafs excepted from life and eftate.

Secondly, all perfons (not being comprehended in any of the former qualifications) who had born command in the

Second clafs to be banifhed, and forfeit two thirds.

the war of Ireland againſt the parliament of England, were to be baniſhed during the pleaſure of the parliament, and to forfeit two thirds of their eſtates; and their wives and children to be aſſigned lands to the value of the other third, where the parliament ſhould appoint.

Third claſs to forfeit a third.

Thirdly, all perſons of the popiſh religion (not being comprehended in any of the former qualifications) who had reſided in the kingdom, at any time from the 1ſt of October, 1641, to the 1ſt of March, 1650, and had not manifeſted their conſtant good affection to the intereſt of the commonwealth of England, were to forfeit one third of their eſtates, and to be aſſigned lands to the value of the other two thirds, where the parliament ſhould appoint.

Fourth claſs to forfeit a fifth.

Fourthly, all other perſons who reſided in Ireland within the time aforeſaid, and had not been in arms for the parliament, or manifeſted their good affections to its intereſt, having an opportunity to do ſo, were to forfeit one fifth of their eſtates.

Ordinance for the ſatisfaction of adventurers and ſoldiers.

By an ordinance made in the year 1653, for the ſatisfaction of the adventurers and ſoldiers, the forfeited lands in the counties of Limerick, Tipperary, and Waterford, in the province of Munſter; the King's and Queen's counties, Eaſt and Weſtmeath, in the province of Leinſter; Down, Antrim, and Armagh, in the province of Ulſter; together with the county of Louth if neceſſary, except the barony of Atherdee, were to be charged with the ſums due to the adventurers and ſoldiers, according to the rates before-mentioned; and to be divided between them by baronies moietively by lot.

And

And for the fatisfaction of the arrears of the forces there, who fhould be immediately difbanded, feveral other proportions of forfeited lands were fet out; particularly, the forfeited lands beginning at the end of one ftatute mile round the town of Sligo, and fo winging upon the coaft, nor above four miles diftant from the fea; which was called the mile-line.

Satisfaction for the difbanded forces in the mile-line.

Purfuant to this ordinance, commiffioners were appointed for putting it in execution, and for taking a furvey of the forfeited lands, and for appointing a court for receiving and hearing claims.

Commiffioners appointed.

And by the inftructions given to the commiffioners, the fecond and third claffes of the Irifh above-mentioned, who forfeited one third or two thirds of their eftates, were to be tranfplanted into the province of Conaught and the county of Clare, for the proportions to be allotted to them, except the mile-line; which line was intended to cut off the communication of the Irifh with the fea, as the Shannon was to cut them off from the reft of the kingdom.

Irifh to be tranfplanted to Conaught and Clare.

Purfuant to thefe ordinances and inftructions, commiffioners of delinquency fat at Athlone, to determine the qualifications of papifts; and upon their decrees other commiffioners who fat at Loughrea fet out the tranfplantations.

Many of the papifts did not take out their decrees, and the tranfplantation was not compleated at the reftoration; although all the papifts lands were feized and fequeftered, and the furveys were in hand and actually taking; and being

Tranfplantation not compleated at the reftoration.

being thus feized and fequeftered on account of the re-bellion, the act of fettlement afterwards vefted them in the Crown.

Thus ftood the fettlement between the parliament of England and the rebels. But, for a clearer explanation of the acts of fettlement and explanation, it is neceffary to take a fhort view of the feveral tranfactions and treaties be-tween the King and the rebels.

Ceffation of arms between the King and rebels.

About the 15th of September, 1643, there was a ceffa-tion of arms agreed upon and declared between the King and the rebels; and the 30th of July, 1646, articles of peace were agreed upon between his Majefty and them, which were afterwards broke by the interpofition of the Pope's nuncio.

Peace con-cluded be-tween them.

Afterwards on the 17th of January, 1648, peace was again concluded between them, which the Earl of Antrim, O'Neill and others in Ulfter, refufed to fubmit to; feveral of thofe who had fubmitted to the King laid down their arms, and upon the general tranfplantion were allotted to their proportions, according to the act for the fettlement of Ireland, in the province of Conaught and Clare. Others of them attended King Charles II. in his exile, after the peace, and waited his reftoration for a reftitution of their eftates. There was alfo another fet of men to be provided for; and thefe were the proteftant officers who had always continued loyal, and had ferved in his Majefty's army, and under his authority, from the beginning of the war to 1649; whofe arrears had never been paid, on account of their loyalty, when Cromwell affigned lands to fatisfy the reft of the army. The King likewife thought himfelf

in

in fome fort obliged to take care of the interefts of thofe adventurers who had lent their money upon the credit of acts of parliament to which his father had affented; and likewife of the officers and foldiers who had lands fet out for fatisfaction of the arrears of their pay.

And, in order to fatisfy all parties he, on the 30th of November, 1660, figned his declaration for the fettlement of Ireland. In which he confirms, in the firft place, to the adventurers all the lands poffeffed by them on the 7th of May, 1659, and allotted to them according to the act of 17 Car. I. as to Englifh or plantation meafure; and engaged to make good the deficiencies of fuch as made proof of them before a certain day.

King Charles IId's declaration for the fettlement of Ireland. Provifion made for the adventurers.

He next confirms the lands poffeffed by the foldiers and allotted them for their pay before the 7th of May, 1659; excepting church lands, and fuch eftates as were either procured by bribery, forgery, or perjury; or fet out by falfe admeafurement; or which belonged to any of the regicides and halberdiers; or to others who had fince his reftoration endeavoured to deftroy the publick peace, or manifefted an averfion to his reftoration and government, or which had been decreed by the court of claims or Exchequer to any perfon.

And for the foldiers.

The officers who had ferved before June 5th, 1649 (except fuch as had received lands for their pay due to them fince that day) were to be fatisfied for their refpective arrears out of forfeited lands in feveral counties therein named.

And for the forty nine officers.

Vol. I.   C c   Proteftants,

And for protestants whose estates had been given to adventurers.

Protestants, whose estates had been given to adventurers or soldiers (except such as had been in rebellion before the cessation, or had taken out decrees for lands in Conaught, or Clare, in recompense of their former estates, were to be forthwith restored, and the others reprised.

And for innocent papists.

Innocent papists, who had been dispossessed, altho' they had sued out decrees and were possessed of lands in Conaught or Clare, in lieu of their former estates, were notwithstanding to be restored to their former estates; and the adventurers or soldiers removed to make room for such papist were to be forthwith reprised. But there was an exception as to innocent papists dispossessed of estates in corporations, who were to be reprised in forfeited lands near such corporations.

And for rebels, who had submitted and adhered to the peace.

Rebels, who had submitted and constantly adhered to the peace, and remaining at home, had sued out decrees and obtained possession of lands in Conaught, were to be bound thereby, and not be relieved against their own act. But if they had served faithfully under his Majesty's ensigns abroad, and had not obtained decrees and lands in Conaught, they were to be restored to their former estates; but not until the adventurer or soldier who was to be removed had a reprise assigned to him, it being more inconvenient to the latter than to the former to wait for reprisals.

Quit rents reserved.

And every such adventurer and soldier so settled, and every person so restored or reprised, was thereby to pay to the King, his heirs and successors, a rent of threepence for every acre in the province of Leinster, twopence

pence farthing for every acre in Munfter, three-half-
pence for every acre in Conaught, and one penny for
every acre in Ulfter, according to the Englifh meafure.

By a commiffion under the great feal of this kingdom,
bearing date 30th of April in the 13th year of his reign,
his Majefty appointed Commiffioners for putting into
execution the feveral matters contained in his faid declaration, and gave them inftructions for that purpofe, by
which they were to caft up the whole debt and demand of
the adventurers, as well thofe who were fatisfied, as
thofe who were in part or the whole deficient, as alfo all
the forfeited lands affigned to or for the adventurers,
according to the furvey commonly called Doctor Petty's
Down admeafurement; and they were to compare the
faid demands and lands together, and what the faid
lands fell fhort of fatisfying the adventurers, according
to the rates, meafures, and proportions, of which all
or any of the adventurers were poffeffed on the 7th
of May 1659, they were to fet apart fo much of the
forfeited lands in the counties of Lowth, Catherlough,
Kildare, or fome other convenient place, for their
fatisfaction.

And in order to the more particular apportioning or
dividing the faid lands among the adventurers, they were
to caufe proclamation to be made in all places in Ireland,
directing every adventurer, his affignee, or agent, who
had received any fatisfaction in land for his adventure,
within 40 days after fuch proclamation, to deliver to the
commiffioners, in writing under his hand and feal, a
certificate of the houfes, lands, &c. poffeffed by him,
together with the content of acres, both profitable and
unprofitable, as the fame were admeafured to him; and

Cc 2   if

if fuch adventure were for houfes in any city, fuch adventurer was to deliver in not only the particulars fet out to him, but alfo the value of them.

And the adventurer: and foldiers to return the furveys of their lands.

And fuch of the faid adventurers and foldiers as had taken furveys of their lands were at a certain day to bring in to the commiffioners fuch furveys, or duplicates thereof, together with the field books ; which the commiffioners were carefully to compare with the furveys taken by order of the late pretended powers ; and if any confiderable difference fhould be found, they were to afcertain fuch adventurer's and foldier's poffeffion, by fuch of the faid furveys as fhould be moft for the King's advantage ; yet fo, that if the foldier or adventurer fhould find himfelf aggrieved, they were to appoint fworn furveyors to re-furvey the lands in queftion, &c. and if information fhould be made to the Lord Lieutenant, &c. that profitable lands were enjoyed for unprofitable, the fame was to be inquired of by a jury ; and the profitable fo enjoyed was to be re-affumed for reprifals of others.

The commiffioners to make up books, &c.

And the faid commiffioners were out of the faid certificates and furveys to make up books of what was due to each adventurer, and to afcertain the poffeffion of fuch to whom lands were affigned, therein expreffing who was the former proprietor, the town-land or denomination, &c. the content, number of acres, the parifh, barony, county, and province in which fuch lands lay ; and where it appeared that any adventurer or his affignee had more lands than were fufficient to fatisfy his debenture, and that fuch perfon was in any other place deficient, or had purchafed the right of any deficient adventurer, fuch overplus was to be affigned unto him towards fatisfaction of fuch deficiencies.

And

And in the restoring of innocents to their estates the
commissioners were to observe the following directions,
viz. not to restore any as an innocent papist, that, at
or before the cessation which was made upon the fifteenth
day of September, 1643, was of the rebels party; nor
any, who being of full age and found memory enjoyed
their estates real and personal in the rebels quarters;
(provided that, where any citizen or inhabitant of the
city of Cork, or of the town of Youghal, or any other
person, was not permitted to live in the English quarters,
but was expelled from thence, and driven into the quarters
of the rebels, then and in such case, such inhabiting
in those quarters, and there receiving any benefit of their
estates, should not be construed or adjudged any bar or
impeachment of their innocence;) nor such as entered
into the Roman catholick confederacy, at any time before
the articles of peace concluded in 1648; nor such as at
any time adhered to the nuncio's or clergy's party, or
papal power, in opposition to the King's authority; nor
such as had been excommunicated for adhering to the
King's authority, and afterwards owned their offences for
so doing, and were relaxed thereupon from their excom-
munication; nor such who derived their titles to their
estates from any who died guilty of any the aforesaid
crimes; nor such as pleaded the articles of peace for their
estates; nor such as, being in the quarters which were
under the authority of the late or present King, held cor-
respondence with, or gave intelligence to, such as were
then in opposition against the late or present King in
Ireland; nor such as before any of the peaces in 1646, or
1648, sat in any of the confederate Roman catholick
                                                assemblies

*The qualifi-
cations of
innocent
rebels.*

affemblies or councils, or acted upon any commiffions or powers derived from them, or any of them; nor fuch as empowered agents or commiffioners to treat with any foreign papal power beyond feas, for bringing into Ireland foreign forces, or were perfons who acted in fuch negotiations; nor fuch perfons as had been wood-kerns or tories before the marquifs of Clanrickard's leaving the government of that kingdom.

Adventurers &c. removed to be reprifed forthwith.

And the faid commiffioners were to take care that adventurers and foldiers in the poffeffion of the eftates of any innocent proteftant, or papift reftored to his eftate, fhould be reprifed in lands of equal value.

To prepare a particular of forfeitures in the counties of Wicklow, &c.

The commiffioners were to prepare a particular of all the forfeited lands in the counties of Wicklow, Longford, Leitrim and Donegal, and alfo of the forfeited lands, &c. not already difpofed of, in Conaught and Clare, being within a mile of the Shannon, or of the Sea, commonly called the mile line, and within any corporation in Ireland; and they were to get the fame valued, deducting the value of the improvements by building or repairing houfes, on any leafes or contracts for leafes in any of the corporations aforefaid, the value of which improvements were to be afcertained.

How to proceed as to the 49 officers.

The commiffioners were then to prepare an account of the perfonal arrears of fuch officers, as ferved in Ireland before the 5th of June, 1649, and had not received any lands or money in fatisfaction of their arrears, before or fince the faid 5th of June, 1649, and the commiffioners were to make an eftimate of the refpective fecurities appointed for the fatisfaction of fuch officers; in order to which,

which, they were to value the houſes, lands, &c. at eight years purchaſe, deducting the value of the improvements, and if the ſaid ſecurity ſhould not extend to ſatisfy twelve ſhillings and ſix pence in every pound of the ſaid arrear, they were to proportion the ſatisfaction according to the ſecurity; then the ſaid ſecurity was to be ſold by publick ſale to the beſt bidder, not under eight years purchaſe, deductions being made for the improvements *. And by theſe inſtructions 18 pence in the pound of the value of the houſes ſet out in the 49 ſecurity, were to be reſerved to the crown.

And for the better quieting and ſettling the ſeveral perſons intended to be provided for by the declaration and this act, the chief governor for the time being, upon certificates ſigned by the commiſſioners, or any five or more of them, expreſſing the names of the perſons, the quality of their eſtates, the number of acres, the barony, county, and province in which ſuch eſtates lay, and the rent reſervable to the Crown, was authorized, at the requeſt of the perſons ſo concerned, to cauſe effectual letters patent under the great ſeal to be paſſed to them, without any further or other orders from the King.

Patents to be paſſed upon certificates ſigned by the commiſſioners &c.

---

* Yet note, that this ſecurity was afterwards made up into lots, and paſt in certificates and patents to certain truſtees, in truſt for the ſeveral perſons concerned in the lots, according to their reſpective debentures, their proportions being mentioned in the patent of every lot; and every perſon concerned had a right in equity to compel the ſaid truſtees or patentees to convey unto him his proportion of the lot, being eſtimated according to the proportion of his debt. But many of the inferior officers have been to this day without ſatisfaction, and the whole has been ſwallowed up by the truſtees, who generally were the principal perſons concerned.

The

Where the
commiſſioners
books were
to be depoſit-
ed.

The commiſſioners, having fully executed their commiſſions, were to deliver up their books unto the Auditor
general, and duplicates of the ſame to the Surveyor general,
to remain there as of record.

Complaints
againſt the
King's decla-
ration.
Carte v. 2.
220.

The King's declaration for the ſettlement of Ireland,
though intended to provide for all intereſts, did not ſatisfy
all parties. The adventurers and Cromwellian ſoldiers had
indeed all that they could aſk granted them therein ; but
the officers who ſerved before 1649, and whoſe loyalty
only had hindered their being paid in the time of the
uſurpation, were treated with great inequality, with regard as well to the quantity of their debt (a proviſion
being only made for the proportion of 12s. 6d. in the
pound) as the ſecurity aſſigned to them, which was not
likely to hold out to anſwer even that proportion. Whatever reaſon they had to complain, their duty and affection
to the King made them declare themſelves ready to be
concluded by his Majeſty's pleaſure. But the Iriſh were
more clamorous ; and thoſe who were entitled to the
benefit of the articles of the peace in 1648 thought it
very ſevere treatment, that their reſtitution ſhould be
poſtponed till reprizals were found out and aſſigned to the
adventurers and ſoldiers who had got poſſeſſion of their
eſtates. They complained ſtill more heavily againſt the
inſtruction given to thoſe commiſſioners, in which the
qualifications for innocency were made ſo very ſtrict.

Reaſon of
one ſevere
mark of de-
linquency.

One of thoſe marks of delinquency, viz. enjoying their
eſtates in the rebels quarters, was certainly rigorous; but
the reaſon upon which it was grounded was, that the
rebellion being almoſt twenty years before, and the Iriſh
having,

having, it was fuppofed murdered all the Englifh, or driven them away, it was not poffible to procure, at that diftance of time, witneffes to prove particular acts of rebellion againft moft of thofe who were therein concerned.

The commiffioners for executing the declaration fat at Dublin, and publifhed proclamations, requiring all adventurers, &c. within forty days to bring in the particulars of their eftates, and all perfons to enter their claims before the firft of May. But very little was done in thefe refpects, for want of a law to warrant the proceedings of the commiffioners; and the judges having declared their opinion, that the declaration, being only an act of ftate, was no warrantable rule to walk by in the difpofing of mens eftates, very few or none of the Irifh entered their claims.

Proceedings of the commiffioners inceffectual. Carte v. 2. 221.

In order therefore to remove fome objections to the King's declaration, and carry it into legal execution, the famous act of fettlement, 14 and 15 Car. II. was made, of which it will be neceffary to give an abftract of the general claufes, with fome obfervations and points adjudged thereon.

Act of fettlement.

And by this act all manors, caftles, houfes, lands, &c. which at any time from and after the 23d day of October 1641, were feized or fequeftered into the hands of, or to the ufe of King Charles I. or II. or otherwife difpofed of, or fet out to any perfon or perfons, ufe or ufes, for adventurers, arrears, reprizals, or otherwife, whereof King Charles I. or II. or any * adventurer, foldier, reprizable perfon

Lands and tenements, &c. feized or fequeftered from and after 23 Oct. 1641, &c.

* Thefe foldiers and adventurers were obliged to claim in order to diveft the lands which by the vefting claufe were in the King; which is plain from the two

fon or others, refpectively, had and received the rents and
profits, by reafon or upon account of the rebellion or war;
or whereof the adventurers, officers and foldiers then or
formerly of the Englifh army in this kingdom, or † tranfplanted or tranfplantable perfons, their heirs or affigns, or
any other perfon whatfoever, upon account of the faid
rebellion or war, were in feifin or poffeffion, on the 7th
day of May 1659; on which were fet apart or referved,
towards the fatisfaction of any the faid adventurers, foldiers,
or other perfons, in confideration of any money or provifions advanced or furnifhed, or for arrears of pay, or in
compenfation of any fervice or other account whatfoever,
or referved in order to a reprizal for fuch incumbrances as
were or fhould be adjudged to any perfons out of the faid
lands, or for any other purpofe whatfoever; or whereof
any *cuftodiam*, leafe for years, or other grant whatfoever,
had been made; or unto which the King's father, or
himfelf were then anywife entitled, upon account of the
faid rebellion or war; or which were then wrongfully detained or concealed by any perfons whatfoever; as alfo

claufes of making out certificates; for by thefe, the commiffioners were empowered
to make out their certificates according to every man's intereft; and on fuch certificates the chief governors, &c. were to order letters patent; fo that the foldier and
adventurer being to begin a title from the King, was to make out his right before
the commiffioners, and the patent was to be granted not in the ufual way, where it
was *ex gratia* by letters from the King, and *fiat* to the chancellor; but this was to
be *ex debito juftitiæ*, and founded only on the certificate of the commiffioners,
without any order from the King. Gilb. rep. 242.

† Tranfplanters were feveral Irifh proprietors of the popifh religion, who, by the
late ufurping powers during the diforder of the times, were difpoffeffed of their
eftates, merely for being papifts; and many of thefe, having afterwards fued out
decrees, were put in poffeffion of lands in the province of Conaught and county of
Clare, in compenfation of their former eftates. Now, in the recital or preamble
of the act of fettlement, it is faid, that tho', as thefe decrees were acts of their own,
they might without any injuftice be denied relief, yet they are reftored fubject to
the provifoes and conditions therein mentioned.

all

all chantries, and all manors, lands, rents, tithes, penfions and other hereditaments whatfoever to them belonging, which were in the feifin or poffeffion of, and out of
which any rent or duty was referved, by any, who by the
qualifications of the act fhould not be adjudged innocent
perfons; as alfo all lands and tenements belonging to any
ecclefiaftical perfons in their politick capacity, and that
had formerly by them been let in fee farm, the right
whereof was in any perfons who fhould be not adjudged innocent; as alfo all leafes that had been made by any ecclefiaftical perfons of any lands or tenements belonging unto
them in their politick capacity, or to any perfons who
fhould not be adjudged innocent, were adjudged and
declared, as from the faid 23d day of October 1641, forfeited to the King his heirs and fucceffors; and were from
that time vefted and fettled in the real and actual poffeffion
of him his heirs and fucceffors; without any office or inquifition thereafter to be found *; notwithftanding the former proprietors or reputed proprietors of the faid eftates,
or any of them, were not attainted for the faid rebellion
or war.

*Declared for-
feited from 23
Oct. 1641.*

But it is declared that the act fhould not extend to the
avoiding of any conveyance or difpofition of forfeited
lands, &c. made fince the 23d day of October 1641, by
any proteftant adventurer, or foldier, or other perfon, of or
from fuch perfons whofe eftates, if they had not fo dif-

*Not to extend
to convey-
ances by pro-
teftants, ad-
venturers or
foldiers, &c.*

* The non obftants in the clofe of this fection was by the Irifh popifh proprietors
thought fevere; but in anfwer to this it was faid, that if there were no attainder of
the Irifh, it was in favour to them; that fo they might not be corrupted in blood,
but left capable to inherit or purchafe afterwards; and that the complaint had been
much more juft if they had been attainted by act of parliament, without further
procefs, as had been done in leffer rebellions. Rep. of Sir Hen. Finch. Carte
v. 2. app. 76.

Dd 2   pofed

poſed of them, would have been confirmed to them by the
rules in the act limited.

Nor to the avoiding of any contract for lands in Conaught
or Clare, ſet out by decrees, made by proteſtants or others
that purchaſed any lands from the perſons tranſplanted
thither; nor to entitle the King to the mean profits of
any of the ſaid forfeited lands, ſince the day aforeſaid ſet
out to any adventurers, ſoldiers, or perſons tranſplanted
into the ſaid province or county ; or let by the late uſurpers
for yearly rents, or granted by them, and confirmed by
the King's declaration aforeſaid, and by the act; other
than ſuch of the ſaid rents then in arrear and unpaid, and
other than forfeited lands concealed.

Nor to be conſtrued to forfeit and veſt in the King any
honours, manors, lands, &c. on the 23d day of October
1641, belonging to the univerſity of Dublin; or to any
archbiſhop, biſhop, dean, prebend, dean and chapter, or
other eccleſiaſtical perſons in their politick capacity; or
any other college, hoſpital, church collegiate or parochial ;
or to the church wardens and pariſhioners of any pariſh
church for the uſes thereof; or to any guild, corporation
or fraternity ; or to any parſon, rector or vicar of any
pariſh church ; or to ſome particular perſons therein named.

Nor to veſt in the King, or take away any eſtate, right,
&c. from any * proteſtants, their heirs, executors, adminiſ-

---

* Theſe innocent proteſtants and papiſts were likewiſe obliged to claim in purſuance of the act, becauſe they were obliged to make out the qualification of innocence ; but when they made that appear, they were not enforced to take out any
new patents, becauſe they were not to begin any new title from the King, but
were remitted to their old title to ſuch lands as they claimed ; and the lands came
out of the Crown, not by any patent, or new grant, but by the diveſting clauſe in
the act. Gilb. rep.

trators

trators or affigns, who did not join with the faid rebels
before the 15th day of September 1643, whereof upon the
faid 23d day of October 1641, they were feized or poffeffed
or entitled; (other than fuch eftate as they were feized or
poffeffed of, to the ufe of the faid rebels) nor to any
judgment or decree obtained by them in the late courts of
claims, or in any of the four courts in Dublin; or for which
any judgment or decree fhould be confirmed or made by
the commiffioners appointed by the King for the execution of his faid declaration and inftructions.

Nor to any eftate, right, &c. of any innocent papift, Nor to inno-
or their innocent heirs, executors, adminiftrators or cent papifts,
affigns.   &c.

All perfons their executors, adminiftrators, and affigns, Perfons to
to whom any lands belonging to fuch proteftant or inno- whom lands
cent papift had been affigned or diftributed, to be firft of innocent
reprized before any other.   papifts were
                                                        affigned, firft
                                                        to be reprized.

All the manors, lands, &c. fo vefted and fettled in To remain in
the King, (except before excepted or provided for as afore- the King to
faid) to remain to the King, his heirs and fucceffors, to clared by the
the intent to be fettled, confirmed, reftored, or difpofed declaration &
to fuch ufes, and in fuch manner, as in and by the faid inftructions.
declaration and inftructions and by the act are declared and
appointed; and the faid declaration and inftructions are
thereby, with the additions and alterations thereby made,
ratified and confirmed.

And the manors, lands, tenements and rents whereof Ecclefiaftical
any archbifhop, bifhop. dean, dean and chapter or any perfons re-
other ecclefiaftical perfons whatfoever, in their politick ftored.
capacity, were actually feized, or poffeffed in the year
                                                    1641,

1641, and through the fury of the times had been difpoffeffed, were to be forthwith reftored, and delivered into
their quiet and peaceable poffeffion.

Saving rights
of others.

Saving the rights of all others (other than fuch perfons
who fhould not be adjudged innocent papifts) by this act.

Leafes for
years by bi-
fhops, &c.
forfeited,
given to their
refpective
fees.

And leafes granted for any certain term of years unexpired, by any archbifhop, &c. or any other ecclefiaftical
perfons, of any lands to them belonging, and which were
by the act forfeited or vefted in the King, are for the remainder of the term unexpired of fuch leafes, given and
confirmed unto the refpective fees or bodies politick to
whom the reverfions belonged; except they lay within the
fecurity of the 49 officers; and except all forfeited leafes,
that exceeded the term of 60 years, of any chantry lands
or houfes lying within the fecurity of the faid officers, and
which were not furrendered nor fentenced to be furrendered to the church in or before the years 1640 or
1641; the remainder of fuch term unexpired being
efteemed part of the fecurity of the faid officers.

Church lands
granted in fee
farm, and for-
feited, to be
allotted by
way of aug-
mentation to
feveral arch-
bifhops, bi-
fhops, &c.
to Trinity
college.

And out of the lands belonging to any archbifhop,
bifhop, &c. or other ecclefiaftical perfons, which had
been granted in fee farm, and were by the act forfeited
and vefted in the King, feveral yearly fums were to be allotted and fet out for the better fupport and maintenance
of feveral archbifhops and bifhops therein named, and their
fucceffors, for ever. And to the provoft of Trinity college,
near Dublin, (out of the forfeited lands in the arch-
bifhoprick of Dublin) and his fucceffors for ever £300 per
annum.

And

And all impropriations or appropriate tithes forfeited or vested in the King, his heirs and fucceffors, by this act, or otherwife forfeited and efcheated to him in right of the Crown (if there were no leafes thereof in being unforfeited, or as foon as fuch leafes fhould expire, or be otherwife determined) are thereby given to the church for ever, and fettled upon the incumbents and their fucceffors, having the actual cure of fouls in thofe parifhes where fuch impropriations were, and fuch impropriate tithes did arife; referving fuch portion thereof to be fettled upon the vicars and choir-men of each cathedral church for the increafe of their maintenance, as the Lieutenant, &c. and council fhould think fit; they the faid incumbents and their fucceffors paying to the King, his heirs and fucceffors, fuch rents and duties as were formerly paid for the fame, with fuch increafe of rents as by the faid Lord Lieutenant and council fhould be adjudged reafonable; or from the expiration of the faid unforfeited leafes refpectively *.

Impropriations or impropriate tithes forfeited to be fettled on incumbents and their fuccellors, where the fame are or do arife.

* For the explanation of the feveral claufes relating to impropriations it is to be obferved, that although the granting words are as full as the claufes about augmentations and the college, yet the feveral bifhops of the diocefes where thofe impropriate tithes lay thought it advifable to pafs patents for them in truft for their clergy; upon which, and not before, the feveral incumbents enjoyed the fame as foon as the old leafes from the Crown expired.

Several rectories being appropriate to religious houfes, they fo continued until the diffolution, and then they came by the King's grant into lay hands, or continued in the Crown, who made long leafes of them to feveral perfons Now, fuch as were the inheritance of the fubject, and were forfeited by this rebellion, the King gives abfolutely by the act to him who had the vicarage, as an augmentation of his living: and fuch as were in leafe to proteftants, or unforfeiting perfons, he grants likewife to the vicars, or thofe who had the cure; and foon after the expiration of the unforfeited leafes, the Duke of Ormond had many of thefe leafes; but as foon as they were expired they came to the clergy, their bifhops having before paffed patents for the ufe of their clergy.

Provided

Lord Lieute-
nant, &c. or
the lord pre-
sidents of
Munster and
Conaught not
to be preju-
diced as to
any impro-
priate recto-
ries, &c.

Provided that nothing in the act should extend to the disposing or altering of any impropriate rectories or tithes, or rents, enjoyed by, or settled on the Lord Lieutenant, &c. or enjoyed by the lords presidents of Munster and Conaught in right of their places.

Saving for
port-corn.

And that the Lord Chief Justice of the King's bench, and Lord Chief Baron of the Exchequer, and master of the rolls, or any other of the King's officers, should and might have and receive such ‡ port-corn of the several rectories which formerly had been paid and reserved.

Lord Lieute-
nant or other
Chief, &c to
allot recom-
pense to re-
storable per-
sons for their
rectories im-
propriate an-
nexed to
churches.

The Lord Lieutenant, &c. to allot such persons (who by the rules of this act should be restored unto the said rectories impropriate, in case no such annexation should be made) such recompense out of the same impropriations as should be thought fit.

Adventurers,
to pay one &c.
year's profit
of their land
to the King.

All adventurers, their heirs and assigns, and all other persons claiming to have any lands or tenements as original adventurers, or under adventurers, were to pay to the King one full year's value of the profits issuing out of the lands possessed and enjoyed as aforesaid, to be paid by two equal payments within the space of two years. And all soldiers, their heirs or assigns, or any claiming under them, were to pay an half year's value of the profits issuing out of the lands possessed and enjoyed by them, in satisfaction of arrears, to be paid at one entire payment.

Erasmus
Smyth, Esq;
the lands set-
tled on him for
any pious uses
exempted
from such
payment.

Provided that all lands settled or conveyed before the first day of May, 1662, on Erasmus Smyth, Esq; for any pious or other charitable use, should be exempted from paying the year's rent herein before imposed §.

‡ For port corn see chap 4.   And

§ This Erasmus Smyth was a considerable adventurer, and concerned in the adventures upon the *doubling ordinance* in 1643, by which all they who adventured money

And forfeited leafes of any meffuages or lands not exceeding 31 years or three lives, from the 20th day of
October, 1641, the immediate reverfion, &c. whereof
belonged to any innocent proteftant or papift, might be
granted by the Lord Lieutenant, &c. unto fuch innocent
reverfioners, who by virtue thereof fhould hold and enjoy
the faid leafes againft the King, his heirs and fucceffors,
and all other perfons.

Refidue of forfeited leafes for 31 years or three lives, where the immediate reverfion was in an innocent proteftant or papift, might be granted by the Lord Lieutenant to fuch reverfioner.

Provided that no undifpofed or unconfirmed lands in
the province of Ulfter, which had come, or fhould come to
the King's hands, fhould be fet out in fatisfaction of deficient adventurers ; but that the fame fhould be wholly referved and difpofed of for reprifal according to the full value
and worth ; unlefs the forfeited lands in other provinces
fhould not be found fufficient to fatisfy thefe deficiencies.

Lands in Ulfter to be referved for reprifals.

money on pretence of carrying on the war in Ireland (though it really was for the
ufe of the parliament in England againft the King) were to have double fatisfaction in
lands in Ireland. Now tho' by this act thefe adventurers were only to have fatisfaction for the fums they really paid, yet they had fo much favour fhown them, that
they might apply the deficiencies of foldiers or adventurers to the overplus, and fo
continue their poffeffion. This Erafmus Smyth, then being an old batchelor, made
the then Duke of York believe that he fhould have the remainder of all his
fortune in cafe he died without iffue, and there being then little probability of his
having iffue, the Duke became agent for him, and got him all the favour poffible
in the act of fettlement. But the pretence was, that feveral publick pious ufes
fhould be performed by the faid Erafmus Smyth after he had paffed patent of his
eftate, which was valued at four or five thoufand a year. The Duke required
him to fettle the eftate according to his promife, which with much ado he fettled on
the Duke in failure of iffue of himfelf, and at the fame time founded a fchool in
Tipperary, another in Drogheda, and a third at Galway, all which were far fhort of
the value of what he propofed to fettle to pious ufes ; however he was connived at
for the reafons before-mentioned ; but he afterwards married and had feveral fons,
and his defcendants have ever fince enjoyed the eftate, fo that the Duke of York
was difappointed, and the intended pious ufes not executed.

E e And

Lands granted by the King, &c. to stand charged with such rents as the lands of adventurers and soldiers.

And it was enacted that all the lands in Ireland granted by the King under his great feal of England or Ireland, and any way ratified by the act, should stand charged with a year's rent, or a year and an half's rent, and such other like quit rents and annual payments wherewith any the lands of adventurers or soldiers stood charged, to be raifed, levied, and paid as other the like rents and payments by the act are appointed to be paid.

Where greater rent was referved in the patent than the quit rent would amount to then fuch rent to be paid, and no other quit rent.

Provided that where a greater rent was referved upon any fuch grants and letters patent than the quit rents referved by this act would amount to, then the rent referved by the letters patent fhall be duly paid, and no other quit rents; faving to the King, his heirs and fucceffors, all right and title to any manors, lands, &c. which he or his father had on the 22d day of October, 1641, in right of his Crown of Ireland; and which were then, or at any time within ten years before, in charge in the Exchequer, (otherwife than by inquifition of lands in Conaught found and returned in the time of the Earl of Strafford's government) and which were not fince difpofed by the King, or his father, by letters patent under the great feal of England or Ireland; and other than fuch rights and titles as in and by a certain act of parliament paffed in England, entitled, An act of free and general pardon, indemnity, and oblivion, were mentioned or intended to be barred or extinguifhed.

Eftates of papifts, reftorable, to be charged with an half year's, if they took no lands in

And the Lord Lieutenant, &c. and council were empowered to charge, for the ufe of the King, the eftates of papifts reftorable by the act, not exceeding the proportions following, viz. all papifts who took no lands in Conaught, one half year's value, and fuch as took

lands

lands in Conaught, one year's value of the eftates unto which they were or fhould be reftored; to be paid into the receipt of the Exchequer for the fatisfying unreftored perfons for want of reprifals; or for the purchafing of reprifals, adventures, arrears, incumbrances, or other allowed interefts by this act, from fuch as fhould be willing to fell their rights; whereby the land defigned for reprifals might the better hold out to anfwer the ends of the King's declaration.

Conaught, and with a year's value if they took lands there, to enlarge the fund for reprifals.

The act of fettlement was far from giving fatisfaction to all parties; it was much complained of by the Irifh, and by none more juftly than the 49 officers, whofe merits did not admit of a difpute, and who were the only perfons that could in ftrict juftice demand the payment of their arrears. They were many of them ancient inhabitants of the kingdom, and of the moft confiderable and beft interefted perfons therein; had loft great eftates by the rebellion, and had diftinguifhed themfelves by their loyalty. Notwithftanding which feveral grants and provifoes had been obtained from the King, and inferted in the act, which intrenched upon the fecurity allotted for the payment of their arrears.

Act of fettlement not fatisfactory, Carte, v. 258.

A bill of explanation was prepared by the Lord Lieutenant and council, and fent over to England in September, 1663; the purport of which was to explain the King's meaning in fome claufes in his declaration, to affign a better fecurity to the 49 officers, to prevent the reftitution of the Irifh to lands and houfes in corporations, (which was done for reafons of ftate, upon a reprefentation from the commiffioners of the multitudes fo reftored) to take away a fixth part of the foldiers and adventurers lands, and thereby to increafe the ftock of reprifals, and to make provifion for

Bill of explanation fent from Ireland, and its purport, Carte, v. 296.

fome

fome eminent perfons who were cut off from all manner of relief by the power of the court of claims being determined. There had been four thoufand claims of innocency entered in that court, yet they had not time to hear fix hundred of them at the day their commiffion ended. The claims of all innocents that had been tranfplanted into Conaught were, by the commiffioners inftruɗions, not to be heard till thofe of innocents who had no land were firft adjudged; fo that not one of them had been heard. There was in the bill fomething done for thefe, and for the people of all forts, towards their fecurity and fettlement, beyond what was done by virtue of the former aɗ. To increafe the ftock of reprifals, the want of which was the great obftruɗion of the fettlement, there was a claufe inferted for refuming into the King's hands all eftates which had been obtained by bribery, forgery, perjury, fubornation, falfe admeafurement, and other undue means, or were enjoyed by perfons that had by any overt aɗ oppofed his majefty's reftoration, or had fince endeavoured the difturbance of the publick peace.

Obj. Aions made to it.

But feveral objeɗions being made on all fides to the bill, all parties grew weary of the unfettled condition in which they found themfelves, and grew difpofed to relax fomething of their feveral pretenfions. To this difpofition in different interefts there were other confiderations which feemed to render a general agreement praɗicable. Upon examination of the pretended aɗs and ordinances of the late times, and the feveral books of fubfcriptions, and the times when they were made, it was difcovered that one entire moiety of the adventurers money was fubfcribed and paid after the *doubling ordinance*, and confequently half of the lands fet out to them ought to be retrenched, they being to receive only fimple fatisfaɗion for fuch money as
                                                            they

they really and *bona fide* advanced. And yet the adventurers who demanded reprifals were moft of this fort.

There were likewife grofs abufes difcovered in the manner of fetting out the adventurers fatisfaction; for they had whole baronies fet out to them in grofs, and then they employed furveyors of their own to make their admeafurements; and thofe finifhed, they had never fince brought in their furveys or field books into the Surveyor general's office, or to publick view. Thus they had admeafured what proportions they thought fit to mete out to themfelves; and what lands they were pleafed to call unprofitable they had returned as fuch, let them be ever fo good and profitable. In the moiety of the ten counties, wherein the fatisfaction of the adventurers was fet out, there were 245,207 acres fo returned. Several perfons, in cafe the King would grant them fee farms of all thofe lands held as unprofitable in five of thofe counties, offered to give an higher rate for them, one with another, than was paid him in quit rents for the profitable; and it was probable that others would do the like in the other five counties. The lands held by the foldiers as unprofitable, and returned as fuch into the Surveyor's office, amounted to 665,670 acres, as appeared by a particular recital thereof in the certificates of the proper officers.

Abufes difcovered in the manner of fetting out the fatisfaction of the adventurers.

Befides the deductions to be made on this account, if the lands both of the adventurers and foldiers were reduced, as they ought, to Englifh meafure, according to the acts, above 500,000 profitable acres would be faved to his Majefty for a fund of reprifals.

When

Propofals of
the Irish
Roman Ca-
tholicks,
Carte, v. 2,
302.

When the bill of explanation was taken into confideration in England, propofals were made by the agents for
the feveral interefts.   The Roman Catholicks, befides a
repeal of the Englifh acts of 17 and 18 Car I. and all attainders fince October, 1641, and the eftablifhment of the
Down admeafurements, and Earl of Strafford's furvey, the
confirmation of all the decrees of the court of claims, and
fatisfaction to poffeffors for improvements, propofed that
all the lands belonging to the Roman Catholicks in 1641
fhould be vefted in his Majefty, who thereout fhould
affign 1,000,000 acres of profitable lands, plantation meafure, to the adventurers and foldiers, in full fatisfaction of
all their pretences to lands for adventures or arrears; to
the 49 officers their fecurity, if it did not exceed their demand; to innocents adjudged by the court of claims, the
lands decreed them; and the reft to be fet out for fatisfaction of publick debts, provifions, &c. before the ceffa-
tion, and to provide for the nominees not already reftored,
in fuch proportions as his Majefty fhould think fit, till
after the fubdivifion and diftribution of the reft among the
feveral interefts; and then one moiety thereof to be applied for the benefit of fuch proteftants, intended to be
provided for in the fettlement, as fhould by the alterations
thereof be moft prejudiced in their refpective fatisfactions,
in fuch proportions as his Majefty fhould think fit; the
other to be diftributed to fuch of the Roman Catholicks
as fhould be the greateft fufferers by the faid alteration.
Thefe moieties were to be taken refpectively out of the
lands affigned to the Irifh and Englifh by the act of fettle-
-ment; each party fupplying what was to be diftributed to
the fufferers of their denomination.   They propofed likewife to advance his Majefty's revenue by quit rents and
                                                        fines

fines payable out of lands to the fum of £200,000 a year; to which all the other interefts readily agreed.

The foldiers defired that the Earl of Strafford's furvey, and Sir William Petty's, might be compared and adjufted, fo as to form thence an authentick rule and ftandard, whereby to know the proprietors quantities, qualities, and values of the lands; and that all the lands whereof the Roman Catholicks were poffeffed in 1641 being vefted in the King, two fifths might be reftored to them, and the other three fifths diftributed among the feveral Englifh interefts. The adventurers complaining that they had loft above 200,000, acres by the decrees of the court of claims, defired that the reft of their fecurity might be continued to them. But thefe propofals being oppofed by the Irifh, all the Englifh interefts agreed in making another, viz. that all the decrees and fettlements already made to the Irifh fhould be confirmed to them, as fully as they were decreed, unlefs the Irifh defired a review; and that 400,000 acres more fhould be fet a part for nominees, but all the reft of the lands vefted in the King to be continued and diftributed among the feveral Englifh interefts. The Irifh objected to the uncertainty of this propofal, as not mentioning where the 400,000 acres were to be fet out, and as being very far fhort of giving fatisfaction to many hundreds of innocents yet unheard, whofe rights were faved by the act of fettlement, as well as to the nominees; computing this deficiency at 170,000 acres, with regard to tranfplanted perfons in Conaught, befides what would be requifite tofatisfy other perfons provided for in the former act, but yet unreftored. This gave occafion to various computa-

tions·

Propofals of foldiers the and adventurers, Carte. v. 2. 303.

tions and difputes about the materials and ftock for reprifals; and to leffen the uncertainty and difficulty of
that matter, the adventurers and foldiers confented to be
reprifed in quantity of acres only, and not in value,
worth, and purchafe.

Final agree-
ment of the
parties, Carte,
v. 2. 303.

The Roman Catholicks at laft to end all difputes propofed, that if, for the fatisfaction of their interefts,
the adventurers and foldiers would part with one third of
the lands refpectively enjoyed by them on May 7th 1659,
in confideration of their adventures and fervice, they
were ready to agree to it; this propofal was in fine accepted, and one third of all the King's grants (except
thofe made to the Duke of York, and fome others) being
likewife retrenched, all matters of any confequence were
thereby adjufted.

And bill of
explanation
agreeable
thereto, Carte
v. 2. 304.

Thus was the fettlement at laft effected by the common confent of all the feveral interefts concerned; and
in confequence thereof the Englifh council, on May 18th
1665, ordered that the adventurers and foldiers fhould
have two thirds of the lands whereof they ftood poffeffed
on May 7th 1659, that the Conaught purchafers fhould
have two thirds of what was in their poffeffion in September 1663, that what any perfon wanted of his two
thirds fhould be fupplied, and whatever he had more
fhould be taken from him; that the adventurers and
foldiers fhould make their election where the overplus
fhould be retrenched; and the 49 men fhould be
entirely eftablifhed in their prefent poffeffions. And
upon thefe refolutions the act of 17 and 18 Car. II. fefs.
5. ch. 2. or act of explanation of the act of fettlement,
was drawn up and received the royal affent.

And

And by this act the lands, tenements, hereditaments, &c. vested in the King by the act of settlement, are declared to be vested in him freed and discharged from all estates tail, and from all conveyances made before the 23d of October 1641, by any tenant in tail, and from all titles and estates derived by, from, or under such conveyance, and from all remainders, reversions, rights, titles, interests, &c. to be disposed of and settled to the uses limited and declared by that and the present act.

Lands, &c. vested in the King by the act of settlement discharged from estates tail, &c.

And it is thereby declared that no person, who by the qualifications in the former act had not been adjudged innocent, should be thereafter adjudged innocent, so as to claim any lands, &c. but should be for ever barred and excluded from all claims, &c.

Persons not already adjudged innocent barred.

And it is enacted that the adventurers and soldiers who on the 7th of May 1659, were seized or possessed of any lands, &c. towards the satisfaction of adventures or arrears, and all deficient adventurers should hold, and enjoy, and be settled in so much of the forfeited lands vested in his Majesty, as should amount to two thirds of what they had, or ought to have had, on the 7th of May 1659, to be computed by Irish measure, according to the Down survey, where the Down survey had been taken, and where the Down survey had not been taken, by the Strafford survey, or by some other survey to be taken; wherein the unprofitable land should be cast in together with the profitable; which two thirds should be held

Adventurers, and soldiers, &c. to be confirmed in two thirds.

Vol. I.   F f   and

and enjoyed by them, in full fatisfaction of any right or claim they might have by virtue of the former act.

Adventurers, foldiers, &c. to be reprifed two thirds, before dif- poffeffed.

And no adventurer, foldier, forty nine officer, or pro-teftant purchafer, in Conaught, or Clare, before the 1ft of September 1663, in poffeffion of lands reftorable was to be removed until he fhould have as much other forfeited lands fet out to him, as fhould amount to two thirds of the lands fo to be reftored.

Directions to be obferved when the ad- venturer, fol- dier, &c. was in poffeffion of more or lefs than his two thirds.

And in order that there might be as little change or alteration of poffeffion as fhould be confiftent with the end of the act, it was directed that where any adventu-rer or foldier fhould be found to have in his poffeffion more lands undecreed away, than his two thirds fhould amount to, he might continue poffeffion of fo much as the commiffioners fhould adjudge his two thirds to amount to, and the overplus to be cut off at his or their election. And the like rule to be obferved in the re-trenchment to be made from the proteftant purchafers in Conaught, and Clare. And where any adventurer or fol-dier fhould be found to be poffeffed of lefs land than his or their two thirds fhould amount to, that then he might continue in poffeffion of what he had, and the refidue to be fet out and made up of other forfeited land, to be allotted and fet out as near as might be to the lands in his poffeffion.

All deficient adventurers to be fatisfied in the fame barony or County.

And it was further directed that all deficient adventu-rers, who were to be fatisfied for two thirds of fuch their deficiencies, and all the adventurers, foldiers, and pro-teftant purchafers in Conaught, and Clare, to whom lands were to be fet out for the making up their two thirds, fhould should be satisfied in the same barony or county, or in the next nearest in value, if it could be conveniently done.

And the commissioners for execution of the act were thereby directed to cause books to be made, in which should be entered the portions allotted to each adventurer, and soldier, &c. for their two third parts; and to return a duplicate thereof into the court of Exchequer, there to remain of record.

Books to be made by the commissioners and returned into the Exchequer.

And upon a certificate under the hands and seals of the major part of the Commissioners, of the lands so allotted, with convenient descriptions and denominations thereof, and presented to the Lord Lieutenant, &c. he the Lord Lieutenant, &c. was thereby authorized and required, to cause letters patent to be passed of such lands, without any further letters or warrants from his Majesty.

Patents to be granted upon certificates of the commissioners.

And where the estate in any lands recovered by any Irish claimant, by any decree by this act confirmed, should not be greater than for the life of such claimant, the commissioners were to give the person against whom the decree should be made his election to take the reversion in fee of such lands, expectant upon the determination of such estate for life, in lieu of his full two thirds; or to have his full two thirds set out to him presently out of some other forfeited lands. And where the estate so recovered by an Irish claimant should be such a remainder, or reversion, as should leave to the person against whom the decree should be made an estate for the life of some other person only, the com-

Where estate for life only recovered by an Irish claimant, the person against whom the decree, &c. might take the reversion in fee or two thirds in other lands. And where an estate for life only left in the person against whom decree, &c. might choose the same in satisfaction of one of the thirds.

missioners

miſſioners were to give the perſon againſt whom the decree ſhould be made his election to continue the poſſeſſion of the land during the life of ſuch perſon, in ſatisfaction of one of his third parts, together with an allotment
of another third part, or to have his full two thirds ſet
out to him out of ſome other forfeited lands.

Such letters patent confirmed againſt the King diſcharged of all eſtates, &c. not therein ſaved.

And all letters patent granted by virtue thereof are
thereby confirmed unto the ſeveral perſons therein named,
according to the eſtates therein granted, againſt the King,
and all other perſons claiming by, from, or under him,
diſcharged of all forfeitures for non-payment of money,
or not putting in of claims, &c. and of all eſtates tail, and
all other eſtates of freehold or inheritance, and all reverſions, remainders, titles and intereſts whatſoever, not
decreed or already allowed, other than what were intended
to be preſerved by this act, and ſhould be reſerved in the
ſaid letters patent.

49 officers confirmed in lands not decreed away, &c.

And it is thereby further enacted, that the proteſtant
officers ſerving before 1649, and not excluded by the
former act, and who had received no lands or money for
their pay due to them for their ſervice, ſhould hold and
enjoy, and be continued and confirmed in all and ſingular
the lands, &c. not already decreed away by the commiſſioners, and in the benefit ariſing from the redemption of
mortgages, ſtatutes and judgments, &c.

Proteſtant purchaſers from tranſplanted perſons to hold two thirds.

And it is thereby enacted, that proteſtant purchaſers
before 1ſt September 1663, from tranſplanted perſons in
Conaught and Clare, ſhould hold and enjoy and be confirmed in two thirds thereof, to be allotted them by the
commiſſioners, and to be entered in books and paſſed by
letters patent, as in the caſe of adventurers and ſoldiers.

And

And it is further enacted, that neither adventurer, foldier, 49 officer, proteftant purchafer in Conaught or Clare, tranfplanted perfon, or any other perfon, entitled to reprifal, fhould be enabled to demand further than 2 full third parts. *No perfon to be repriled above two thirds.*

And it is further enacted, that all lands by this or the former act vefted in the King, or reftored by virtue of any decrees, or by virtue of any claufe in this or the former act, and not particularly, by exprefs words, excepted from quit rents in the fame claufe, fhould be fubject and liable to fuch quit rents, to be paid his Majefty, as in the former act is directed * ; faving only that the lands in the province of Ulfter, which by the former act were charged with one penny the acre quit rent, fhould be thenceforth charged with two pence the acre quit rent. *All lands vefted in the King, or reftored, fubject to quit rent as in the former act.*

Lands in Ulfter to two pence the acre.

But a power is thereby given to the Lord Lieutenant and council, within the fpace of three years, to make fuch moderation or abatement of the quit rent as they fhould think fit, where the quit rent fhould be fo near the value of the land as to difcourage improvement, which order of council, enrolled in the Exchequer, is thereby made as effectual as if thereby enacted. *Power given to the council to abate quit rents.*

And decrees made by the commiffioners under the former act, whereby any proteftants had been declared innocent, are thereby confirmed. And † decrees whereby *Decrees of innocency confirmed.*

* On this claufe chiefly it was determined in this court, in the cafe of the King v. Dardis, Hill. 1667, that the eftates of innocent papifts fhould be fubject to quit rents.

† The decree of the commiffioners could not reduce the eftate of an innocent papift, fo as to give him a lefs eftate than what he had before, for the decree did not make a title to the innocent papift as it did to the adventurer or foldier. Gilb. rep. Kellet v. Mc.Carty Moore.

any

any papifts had been declared innocent, and which fhould be taken out within a certain time, are thereby likewife confirmed, with fome exceptions and reftrictions; but the perfons fo declared innocent and reftored are thereby debarred from fuing for mean profits.

Decrees of innocency *quoad hoc*, not to entitle claimants to any other lands.

And whereas many perfons had put in their claims before the former commiffioners, as innocent perfons, thereby demanding fome fmall parcel of land only, or deriving a title to fome fmall part from fome Irifh papift, and thereupon no oppofition being made, the commiffioners declared the claimant, or the perfon under whom the claimant derived, to be innocent *quoad hoc*; after which the claimants, &c. alleging themfelves to be declared innocent, entered upon great eftates in feveral counties, as divefted out of the Crown by fuch judgment of innocence, whereas if the whole eftates, to which the claimants pretended, had been then in queftion before the commiffioners, the adventurers and foldiers therein concerned would have been fummoned, and might have produced proof of their nocency; it is therefore enacted, that no fuch decree of innocency *quoad hoc* fhould give fuch perfon any title to enter upon or enjoy any other lands, than what were particularly mentioned in fuch decree.

Innocents left to law, the proceedings againft them.

And becaufe feveral perfons had been decreed innocent, but had neverthelefs not been reftored, but had been left to the courfe of * law for the recovery of their poffeffions, by trying their titles, it is thereby enacted, that the defend-

---

* This was where a papift was innocent, and pretended a title to land, and fummoned the proprietor to appear before the commiffioners, and the proprietor not only denied the innocence but likewife the title of the papift claiming the eftate, there if his innocency was found, he was found innocent at large and left to the law to try his title to the eftate.

ants

ants in such claims should within three months make
their election, whether they would relinquish the possession of the lands in controversy to the King, and resort
to other forfeited lands for their two thirds, &c. or abide
a trial at law; and if they should elect a trial, and the Irish
claimant should fail to prosecute his title, or a verdict
and judgment should pass against him, then the adventurer, soldier, &c. was to hold the land to him and his heirs;
but if a verdict and judgment should pass for the claimant,
or no such election should be made, then the adventurer,
soldier, &c. was to be excluded from demanding his two
thirds. But it is thereby directed that no other title should
be given in evidence by such Irish claimant, but such as
was alleged in the claim exhibited before the commissioners.

And the act further directs, that in case of doubts or
defects arising or appearing therein, the commissioners
might, within two years next after their first setting, acquaint
the Lord Lieutenant and council therewith, and that such
order of amendment or explanation as they should make
in writing, within the said two years, and enrolled in
Chancery, should be as effectual as if it had been part of
the act.

Doubts
arising to be
explained by
Lord Lieute-
nant and
council.

And in consequence of the last mentioned clause in the
foregoing act, and of certain doubts proposed to them by
the commissioners, they did by an order of council, bearing
date 9 April 1666 and enrolled in the court of Chancery,
order and declare, first, that all estates, &c. which did on the
23d of October 1641, or at any time since, belong to any
Irish papist, or which had been returned by the civil survey, or Down survey, as belonging to any Irish papist, and
which at any time after the 23d of October 1641 were
seized,

Resolution of
doubts by
Lord Lieute-
nant and
council.

Estates seized,
sequestered,
&c. to be
taken and ad-
judged for-
feited to and
vested in his
Majesty,
without fur-
ther proof.

feized, or fequeftered, or vefted in his Majefty, upon account
of the rebellion (excepting fuch eftates as had been decreed
to innocents, and belonged to them on the 22d of October
1641, and excepting fuch lands as had been reftored to
the former proprietors by fome claufe in either of thofe
acts, and excepting any lands for which fome judgment
or decree was had by a proteftant in the late court, or
pretended court of claims, or in any of the four courts,
before the 22d day of Auguft 1663) fhould at all times
thereafter in the four courts fitting in Dublin, and in all
courts of juftice, and in all trials, actions and fuits, both
in law and equity, as well between his Majefty and any
of his fubjects, as between party and party, without any
further proof, be always conftrued to have been feized,
fequeftered, and from the 23d of October 1641 forfeited to
his Majefty, without any inquifition or office found, &c.

After adjudi-
cation of any
lands, &c. and
after certifi-
cate and pa-
tent paffed,
the rights of
all perfons,
(except as
herein are ex-
cepted) for
ever con-
cluded and
barred,

And fecondly, that after the commiffioners for executing
the faid acts fhall have adjudged any of the faid lands, fo
vefted or forfeited to his Majefty, to any perfon or perfons
who by the faid acts are * entitled thereunto, and fhall have
granted their certificate accordingly, and letters patent
fhall be thereon paffed, the rights, titles, and interefts of
all perfons whatfoever who had not been adjudged innocent, as well fuch as were proteftants as papifts, fhould
be thereby concluded and barred for ever; other than
fuch rights and titles which fhould be referved in the
letters patent; and other than fuch rights as are the proper
act of the party, to whom fuch letters patent fhall be
granted, or of thofe under whom he claims as heir,
executor, or adminiftrator; and other than fuch debts,

---

* By *the perfons entitled* muft be underftood perfons entitled by the act to receive
certificates. Gilb. ante 216.

leafes,

leafes, and payments whereunto the fame are by the faid
acts made liable; and that the faid lands, &c. in the faid
letters patent contained, fhould be by the faid acts confirmed, according to the feveral eftates thereby granted,
againft the King and all other perfons bodies politick and
corporate.

And thirdly, that all adventurers and foldiers, their heirs
and affigns, fhould have and enjoy an eftate of inheritance
in fee fimple in fuch lands as fhould be certified to belong
to them; unlefs fome leffer eftate fhould be therein exprefsly limited; and that in cafe fuch leffer eftate fhould
be fo limited, the party fhould be reprifed out of other
lands, fo as to make up his two third parts, by the faid
acts intended to him, equal in worth and value to others
who fhould have eftates in fee fimple certified and granted
to them.

Upon the conftruction of thefe acts, a confiderable
queftion arofe, whether the eftates of innocent papifts
reftored were liable to quit rents. When the fettlement
of Ireland was under confideration before King Char. II.
previous to the paffing the declaration, the Irifh papifts
urged that they fhould only hold by their old tenures, and
pay their old rents; but the agents on the other fide
(amongft whom was the earl of Orrery) infifted that as the
rebellion was begun by the papifts, whereby the King was
fo long deprived of his ancient revenue, the papifts ought
to contribute equally with the new interefts for its future
augmentation. After the declaration, and when the act of
fettlement was before the council, the papifts again urged
this point; but the commiffioners from the lords juftices
anfwered that this matter had been fettled by the declaration, and the former arguments were ufed in fupport of

Vol. I. Gg the

What eftate the adventurers and fol-diers fhall have in the lands fo ad-judged, &c.

Determina-tions relative to the quef-tion whether eftates of in-nocent pa-pifts were liable to qui: rents.

the charge, and the bill paffed without alteration as to this point.

But when all the decrees of innocence were paffed under the act of fettlement, and the eflates put in charge for the quit rents, it was made matter of doubt, whether under that act they were liable. And the point was brought under confideration of the court of Exchequer in Eafter term, 1665, in the cafe of the King againft Gerald Dardis, which was as follows:

He, being charged for fome lands in the county of Weftmeath, came in and pleaded his decree of innocence; the Attorney general replied, that he would not profecute further, and did not deny the lands being put out of charge, with a faving for the arrears before the 15th day of June, 1663; the court accordingly gave judgment of exoneration: And the lands of innocent papifts were taken out of charge, and fo continued until after the act of explanation *. When that act paffed they were again put in charge; and the point came again into queftion in Hillary term, 1667, in the cafe of the King againft the fame Gerald Dardis, as follows:

After praying oyer of the charge, Dardis pleaded, that before the charge, viz. 22d of October, 1641, he was feized of thefe lands in his demefne as of fee; that being fo feized, 1ft of May, 1652, the lands were feized and fequeftered; he then pleaded the act of fettlement, and the exception in favour of innocent papifts; and his decree of innocency, and the former difcharge, and the act of ex-

---

* Thofe rents amounted at this time to £10,000, as appears by a letter of the Duke of Ormond to King Charles II. Carte, 2 vol. app. p. 87.

planation,

planation, confirming the decrees of innocency. The Attorney general replied, admitting the several matters so pleaded, that the claufe in the act of explanation by which all lands vested in the King by, or restored by virtue of, any decree, or by virtue of any claufe in either act, and not particularly excepted from quit rents in the fame claufe, should be liable to quit rents, and averred, that the lands in queftion are not exempted by exprefs words, and that the rent in queftion was a quit rent, according to the rule in the act of fettlement. Dardis demurred; and the Attorney general joined in demurrer; and after many arguments, and great deliberation, the court gave judgment for the charge: for that the lands of innocent papifts were vested in the King, and restored by virtue of decrees, and thofe decrees confirmed by the act of explanation, and confequently they fell within that claufe in the act which charges quit rents; and there are not in either act any exprefs words to exempt them, as there are with regard to innocent proteftants.

And afterwards, Michaelmas term, 1670, upon motion made by John Temple, knight, fetting forth, that in the roll tranfmitted by his Majefty's late commiffioners for executing the acts of fettlement and explanation, for the charging the eftates of innocent papifts, there were many lands left out of charge, and others unduly charged with more acres than were in the Down furvey; and therefore defiring that fuch lands fo omitted out of the roll, and included in the refpective decrees of the faid innocent papifts, might be brought in charge, by his Majefty's Surveyor general, and that where any fuch miftakes fhould be in the roll of innocency the fame might in like manner be certified. It was ordered that his Majefty's Auditor general fhould, upon certificate of the Surveyor general, bring in

Michaelmas term, 1670, ordered that the lands of innocent papifts fhould be put in charge.

G g 2

charge

charge all fuch lands fo decreed to any innocent papift, and
fo omitted out of the faid roll in charge ; and that he likewife, upon the like certificate from the Surveyor general,
fhould afcertain where any miftake fhould be in the faid
roll, fo that his Majefty's rents might be thereby afcertained;
whereof the Auditor general, Surveyor general, and all
other officers of the court were to take notice.

Innocent pro-
teftants de-
riving under
innocent pa-
pifts liable to
quit rent.
    And on the 13th of December, 1673, in the cafe of the
King againft Malone, the queftion being, whether an innocent proteftant deriving under an innocent papift fhould
be liable to quit rent, the court were of opinion that he
was, and gave judgment for the King.

    By the ftatutes of 10 Will. III. c. 7, 10 Will. III. c. 18,
and 2 Ann, fefs. 1. c. 9. there are feveral provifions made
for quieting poffeffions under the acts of fettlement and
explanation, and barring ancient claims ; but thefe acts,
being now of little ufe, are not neceffary to be inferted
here.

C H A P.

C H A P.   XXII.

Of the FORFEITURES in this KINGDOM by the REBELLION in 1688.

VERY fhortly after their Majeftics, King William and
Queen Mary, had accepted the Crown of thefe
realms, which was on the 13th of February, 1688, a rebellion broke out in this kingdom in favour of the late
King James II. encouraged and affifted by the French King;
which, after it had raged for near three years, was quelled,
and the Irifh reduced to obedience, at the expenfe of the
people of England; for which reafon, and as the forfeitures
were very confiderable, (many perfons of this kingdom of
large properties having been engaged in the rebellion) the
parliament of England, notwithftanding the royal prerogative, and the right claimed by the Crown to the
difpofal of thefe forfeitures, took upon them to difpofe of
them as they thought fit; and even to re-affume almoft
all the grants which the Crown, in virtue of this prerogative and right, had made to feveral perfons of feveral of
thefe forfeitures.

The rebellion in 1688.

Now the ftate of thefe forfeitures upon this rebellion,
and the manner in which they were difpofed of, were as
follows, as appears by a report made by the commiffioners
of the revenue to the lords juftices the 3d day of June,
1693.

The ftate of the forfeit-
ures, and how difpofed of.

Soon

Commission to the Earl of Longford and others.

Soon after the reduction of Ireland, their Majefties granted a commiffion to the Earl of Longford, and others, for feizing and fecuring all forfeited goods, chattels, and eftates, dated the 12th of July, 1690.

And reprefentation againft it by the commiffioners of the revenue.

Upon which the commiffioners of the revenue, by their letter of the 17th of July, 1690, to Sir Robert Southwell, reprefented to his Majefty that they were empowered by their commiffion, as well as by particular directions of the lords of the treafury, to take care of the forfeited eftates and effects belonging to rebels in this kingdom; and that the management thereof by them, and their collectors, within their refpective diftricts, would be much more effectual and lefs chargeable than by others.

His Majefty's order thereon, by which the commiffioners of the feizures were to feize and to tranfmit to the commiffioners of the revenue.

His Majefty by his order of the 23d of July, 1690, fignified his pleafure, that the faid commiffioners of feizures fhould neverthelefs continue to act by virtue of their commiffion, but that all feizures that were made by them fhould be tranfmitted to the commiffioners of the revenue, to the end that fuch forfeited goods as were perifhable might be difpofed of; and that the houfes and lands might be fet for a year by the faid commiffioners of the revenue; the produce thereof to be paid to their Majefties receiver general.

And feveral returns accordingly tranfmitted by them to the commiffioners of the revenue.

And purfuant to this order, and to a fubfequent order of the lords juftices, dated the 29th of September, 1690, the commiffioners of feizures did tranfmit to the commiffioners of the revenue the returns of lands and goods feized by their fub-commiffioners, contained in feveral fchedules or lifts; which the commiffioners of feizures certified to be true copies of the returns made to them by

their

their fub-commiffioners. But this was not done till the months of October, November, December, 1689, and January, 1690, as appears by the dates of the feveral tranf-mits figned by the commiffioners of feizures.

Thefe fchedules being fo tranfmitted, the commiffioners of the revenue caufed fuch of them as related to perfonal eftates to be tranfcribed, and fent to the feveral collectors in whofe diftricts fuch perfonal eftates were, with inftructions, which were approved of by the lords juftices, to them to demand and receive the goods, ftock, &c. therein mentioned from the fub-commiffioners, (giving acquittances for the fame) and alfo to difpofe of them, when received, as therein directed.

The commiffioners of feizures being fuperfeded by warrant under the great feal, dated the 6th day of February, 1690, it was then thought neceffary for their Majefties fervice, that the original returns of the fub-commiffioners, then in the hands of the diffolved commiffioners, fhould be lodged in the Chief remembrancer's office in the Exchequer, there to remain on record, as a check upon all fuch perfons as had been concerned in the forfeitures; which was done accordingly.

And purfuant to the before-mentioned inftructions the feveral collectors demanded from the feveral fub-commiffioners the goods, corn, ftock, &c. wherewith they were charged in their faid fchedules, and made inventories of fuch part thereof as they received; which, with the accounts of their proceedings from time to time, they returned to the commiffioners of the revenue.

But

But though the returns made by the fub-commiffioners
were very large, and carried with them an appearance of
confiderable quantities of forfeited goods, &c. yet by the
reafons entered in the margins of the faid returns by the
fub-commiffioners themfelves, it appeared that but a fmall
part of them could be expected to be received by the collectors; thefe reafons fetting forth that the goods fpecified
were either claimed by perfons under protection, or detained by proteftant landlords for rent due to them from
forfeited tenants, or were feized on and embezzled, or deftroyed by the army.

The commiffioners of the revenue finding by the returns of their collectors, and by other informations, that
there neverthelefs remained with feveral of the fub-commiffioners confiderable quantities of forfeited goods, flock,
&c. which they ought to have delivered or accounted for,
did in Michaelmas term, 1691, confult with their Majefties council, and foon afterwards with the Barons of the
Exchequer, what method would be moft proper to bring
the fub-commiffioners to a full and particular account;
upon which it was refolved that perfonal interrogatories
fhould be exhibited to them, and that upon perufal of
their anfwers fuch further profecution fhould be made as
the court fhould think fit.

Accordingly, interrogatories were on the 10th day of
February, 1691, filed, and feveral of the fub-commiffioners, and perfons employed under them, examined
thereon; but upon perufal of their anfwers the King's
council found them fhort and evafive, and that there was
reafon to proceed againft fome of them in another method,
which it was refolved fhould be by informations in the
Exchequer;

Exchequer, and the profecution was preparing and carrying on, when the late commiffioners of the revenue were fuperfeded in Auguft 1692.

When the revenue was committed to the management of the commiffioners, it was declared at the fame time, that there would be very foon a parliament in Ireland; and the commiffioners had fcarce entered upon the reft of the bufinefs, and began to inquire into the nature and condition of thefe forfeitures, when (the elections being over) it appeared that feveral of the commiffioners and fub-commiffioners were chofen members of the houfe of Commons; and the commiffioners believing it not fit for them to give any trouble to the members at the time of their fitting, they concluded it beft to refpite all proceedings of that kind till the rifing of the parliament. *The reafon.*

And upon the prorogation of the parliament there iffued immediately a new commiffion to inquire exprefsly into the perfonal forfeitures, which in that branch fuperfeded that to the commiffioners of the revenue, fo that for thefe reafons they did not at all intermeddle in the faid matter. *A new committee to inquire into perfonal forfeitures.*

This is the fum of the proceedings between the commiffioners and fub-commiffioners of forfeitures, and the commiffioners of the revenue, concerning thefe forfeitures.

Now the manner in which the collectors accounted for the fame to the commiffioners of the revenue was as follows. *The manner in which the collectors accounted to the commiffioners of the revenue.*

As to all the goods, corn, ftock, &c. which came to any of the collectors hands, they at several times returned up particular accounts thereof, in the charge part of which they made themfelves debtors, according to the feveral inventories to the faid accounts annexed, for all the goods, &c. which came to their hands, whether the fame were received by them from the fub-commiflioners, or feized by themfelves, or received from the commiffaries general of provifions by orders of the govern-

ment; and in order to afcertain the faid charge the better upon them, they were required to make affidavit that their faid accounts contained all the goods, &c. that had refpectively come to their hands.

The difcharge parts of the faid accounts contained the manner how the faid goods in particular were difpofed of under the following heads, viz.

Goods delivered, by orders of the government, or by orders of the court of Exchequer, and the late commiffioners, to perfons that made out a right to the fame; eftimated at about £5000.

Bread, corn, hay, and oats, and other provifions, &c. delivered by order of the government to the commiffaries general of provifions.

Bullocks, oxen, or horfes, fit for carriage, or draught, delivered by like order to William Robinfon, and Francis Cuffe, Efquires, for the ufe of the train of artillery.

Goods,

Goods, &c. fold by publick cant, purfuant to the inftructions before mentioned; as to which the collectors
were required to make oath, that the fame were fold for
the particular rates charged in their accounts, for their
Majefties beft advantage, without any private benefit to
themfelves.

Goods remaining undifpofed, being for the moft part
lumber, and goods of fmall value, which the collectors by order of the commiffioners of the revenue delivered to the commiffioners of infpection.

Several of thefe accounts paffed upon oath in the Exchequer, as the refpective collectors could be fpared to
come up to pafs their general accounts.

The commiffioners of the revenue alfo received from
the commiffioners of forfeitures feveral bonds, taken by
them or their fub-commiffioners, amounting to £18290,
fome whereof were from proteftants in poffeffion of lands
by mortgage from the forfeiting proprietors, amounting
to £5900. The condition of the bonds was to account to
their Majefties for the overplus profits of fuch lands; but
by the calamity of the times, the lands were not found
worth the intereft of the money; others were from proteftants not forfeiting, or papifts under protection, laying
claim to goods which had been feized, amounting to
£12590; the condition to anfwer the value of the goods
therein mentioned, if their Majefties title fhould be made
out in a fhort time limited; but this condition putting
the proof upon the King made the bonds of little or no

Several bonds alfo delivered by the com-miffioners of the forfeit-ures to the commiffio-ners of the revenue and how dif-pofed of.

H h 2   value;

value; fome few of thefe bonds were lodged in the Exchequer in order to profecution, the reft were delivered
to the commiffioners of infpection as aforefaid.

The goods
for which the
bonds were
taken alfo
returned as a
charge.

It is to be obferved, that all the goods for which the
faid bonds were taken were likewife returned as a
charge by the late commiffioners of forfeitures in their
fchedules of forfeited goods, which fwelled their accounts
by two charges for the fame thing, amounting each to
the fum of £12390.

Proceedings
of the com-
miffioners of
the revenue
as to real
eftates.

Having thus given an account of the proceedings of
the late commiffioners of the revenue concerning the
perfonal eftates forfeited to their Majefties, it now remains to give an account alfo of the real eftates.

Order to fet
the lands for
one year.

The order before mentioned from the lords juftices, of
the 19th of September, 1690, which directed the commiffioners of forfeitures, to deliver to the commiffioners of the revenue lifts or fchedules of all the lands
feized by them or their fub commiffioners, did alfo direct
the commiffioners of the revenue to fet the fame for one
year, for their Majefties beft advantage, purfuant to
certain methods propofed to the lords juftices, and approved of by them.

Further order
to fet the
lands for one
year.

The lifts or fchedules of forfeited eftates, which were
firft delivered to the faid commiffioners, appearing to them
to be very faulty and deficient, in not returning lands of
perfons forfeiting, and in returning lands of perfons not
forfeiting, upon reprefentation thereof to the lords juftices, their lordfhips did by their order of the 7th of
October, 1690, direct the commiffioners of the revenue

to

to fet all fuch lands for one year as fhould appear to them by information, or otherwife, to be forfeited to their Majefties, not tieing themfelves up to the returns of the commiffioners only.

But in fome cafes, where the faid commiffioners perceived a combination among the bidders for fome of the faid lands to be fet, they fometimes adjourned fetting the fame till a further day, whereof they then ordered further publick notice, to procure more bidders for the lands; particularly in the cafe of fome baronies of the Earl of Antrim's eftate, and other lands.

Adjournment of Setting, where combination was fufpected.

And fometimes they received fpecial orders from their Majefties and the government, to fet particular lands at a certain rent without canting, the yearly value of which at the time they were fet, amounted to about the fum of £7428, tho' actually fet but for £5571.

Some lands fet by fpecial orders of their Majefties, not by cant.

Thefe orders were granted, either purfuant to a claufe in his Majefty's declaration, for perfons that would come under his protection, wherein it is declared that they fhould be allowed out of their forfeited eftates a proportion for their maintenance according to their qualities, or elfe for fome fervices done.

Purfuant to his Majefty's declaration or for fome fervices done.

But as well for the year 1691, as the year 1692, there were feveral parcels of lands pofted by the commiffioners to be fet as aforefaid, for which no bidders appeared; of thefe lands the commiffioners caufed lifts to be drawn out, and fent them to the refpective collectors in whofe diftricts they lay, to fet them for one year for their beft advantage, which the collectors did, as to

How the commiffioners proceeded where no bidders appeared.

fuch

such as they could get tenants for; but in several counties,
the country being so full of rapparees, the improvements for the most part destroyed, and the lands waste,
(particularly in the counties of Longford, Limerick,
Tipperary, and in the most part of Connaught, &c.) no
tenants for one year could be had for them: but for such
of the said lands as were set by the collectors, or inhabited
by any tenants, the collectors charged themselves with the
produce thereof, in their accounts upon oath in the
Exchequer.

The commif-
fioners of the
revenue in-
creafed the
rents accord-
ing to the corn
and fallow.

The commissioners of the revenue, the first year they
set the said lands, took care to increase the rents thereof,
according to the number of acres forfeited, and the corn
and fallow that appeared to them to be on each parcel of
land, by the proposals of the persons bidding, or by other
informations; and added a clause in the leases, that if
there appeared any more corn or fallow than what was
valued and included in the rent of each lease, the lessee
should pay at the rate of 20s. per acre of corn, and 5s.
per acre for fallow, for such overplus, and the collectors
had directions to inquire and return in their accounts,
where they found any such overplus; but the greatest
part of the corn and fallow returned by the sub-commiffi-
oners of feizures, to be sown, or made on the forfeited
lands, did appear not to be forfeited, but to belong to the
under tenants, who sowed and made the same, and who
were generally either protestants, or papists under protection, who could pay no more than the rent reserved on
them, where they had leases from the forfeiting persons,
or the custom of the country for the standing thereof,
where they had none.

In

In the leafes made by the commiffioners, there was a clanfe that the rent therein referved fhould be paid to their Majeflies, clear, over and above all taxes, charges, &c. whatfoever; whereas other landlords did allow their tenants the militia money, and other extraordinary charges which at that time lay heavy upon the country.

And referred the rents clear over and above all taxes.

And the commiffioners did in feveral cafes (when they could) oblige the tenants that took the faid lands, the firft year, to be accountable for the arrear due thereout before they took them; and where they could not, they gave it in their inftructions to levy, from thofe that enjoyed the faid lands before they were fet, all fuch arrears, or fo much thereof as could be got, which the collectors in their accounts on oath charged themfelves with.

And obliged the tenants to be account-able for the arrears due before they took.

Laftly, feveral of the leafes made of thefe lands by the commiffioners having determined the firft of November, 1692, and the commiffioners having informed the then chief governor thereof, he ordered them to fet the fame for three years; but when they were going to proceed thereon, he countermanded the order, directing them to give notice to the faid tenants, that they fhould refpectively continue to hold the lands for the half year ending at May following, upon giving fecurity for payment for the faid time, after the rate they paid by their expired leafes; to the end that all the leafes that fhould thenceforth be made of the forfeited lands might commence from May day, 1693, which they did accordingly by publick notice.

New order for fetting the lands.

But

But before the Commissioners of the revenue set them, a commission issued to inspect and inquire into the forfeited lands.

But before the time came for their making such leases, a new commission issued to commissioners, empowering them to inspect and inquire into the value and management of all the said forfeitures, and to set all the forfeited lands, with many other powers; as may appear by the enrolment of the said commission in the rolls office of this kingdom *.

Differences between the King and house of Commons concerning these forfeitures.

But great differences soon after arose between his Majesty and the English house of Commons, concerning these forfeitures; it having been resolved by them that a bill should be brought in, for attainting the persons who had been in rebellion in England and Ireland, and for confiscating their estates, and applying the same to bear the charges of the war, reserving to the King a power to dispose only of a third part of them; which was considered by the court party as a violation of the right of the Crown, for that his Majesty had an undoubted right, in virtue of the prerogative, to dispose of these forfeitures as he should think proper.

The King grants them away as he thought fit.

However, as this bill was likely to lie long before the lords, many petitions having been offered against it, the King, in order to bring the session to a speedy conclusion, had promised that the matter should be kept entire until the next session; which passing away without any proceeding in it, his Majesty thereupon granted away all these confiscations as he thought fit.

* There are three of these commissioners enrolled in the rolls office of the following dates to wit, 12 November, 4 Will. III. 29 March, 7 do. and 24 February, 8 do.

It

It was then immediately alleged that these forfeitures would yield a million and a half in value. Great objections were made to the merits of some who had the largest share in those grants. Attempts had been made in the Irish parliament to obtain a confirmation of them; but the earl of Athlone's only was confirmed; so that it became a popular subject of declamation to arraign both the grants and those who had them. Motions had been often made for a general reassumption of all grants made in this reign; to which it was answered by the court party, that since no such motion was made for the reassumption of those made in the reign of King Charles II, notwithstanding the extraordinary profusion of them, and the ill grounds upon which they were obtained, it showed both a disrespect and ingratitude, if, while no other grants were reassumed, this King's only should be called in question; and they proposed, that if the retrospect were carried back to the year 1660, they would consent to it, and urged that what would arise by such a retrospect would be worth while. But the infinite perplexity that would be occasioned by the unravelling, after such a length of time, the many sales, mortgages and settlements, which had been made pursuant to those grants, was an unanswerable objection to this proposal.

Debates in the House concerning these grants, and a reassumption thereof.

But at length a more effectual method was taken; for in the 10th and 11th years of his Majesty's reign an act of parliament passed in England, whereby a commission was given to seven persons named by the Commons, to inquire into the value of the forfeited estates so granted away, and into the considerations upon which these grants were made.

An act by which a commission is granted to seven persons to inquire into these forfeitures.

Vol. I.   I i   Accordingly

Zeal of the commissioners in exaggerating the value of the grants, and depreciating the merit of the grantees.

Report delivered to the house by four only.

Accordingly thefe commiffioners, namely, the earl of Drogheda, Francis Annefly, John Trenchard, James Hamilton, Henry Langford, Sir Richard Levinge and Sir Francis Brewfter proceeded in the execution of this commiffion, in which they fhowed that out of the fale of the confifcated eftates £1,699,343 might be raifed. They difagreed in fome points, which caufed the report to be delivered to the houfe by four only of the feven commiffioners; the other three, namely, the earl of Drogheda, Sir Richard Levinge and Sir Francis Brewfter, refufing to fign it, thinking it falfe and ill grounded in feveral particulars, of which they fent an account to both Houfes; but no regard was paid to their memorial, nor any inquiry made into their objections; the fpecious propofal of raifing fuch a large fum towards difcharging the publick debts prevailed fo with the houfe, that no complaints againft the proceedings of the commiffioners could find admittance, and all the methods ufed to difgrace the report had the contrary effect *.

The

* The report confifted of ninety articles, the chief of which are thefe.

The number of acres in the feveral counties belonging to forfeiting perfons. } 1,060,792

Which being worth £211,623 a year, at fix years purchafe for a life, and at 13 years for an inheritance, amounted to } £2,685,130

Out of thefe lands, the eftates reftored to the old proprietors by the articles of Limerick and Galway are valued at £724,923, and thofe reftored by royal favour at £260,863, after which and feveral other allowances, the grofs value of the eftates forfeited fince the 13th of February 1688 amount to } £1,699,343

The No. of grants and cuftodiams fince the battle of the Boyne under the great feal of England are 76, fome of the principal of which are mentioned, viz.

	Acres.
To the lord Romney 3 grants of	49,517
To the earl of Albemarle 2 grants of	108,633

To

The Commons, having examined this report, came to an unanimous refolution, that a bill fhould be brought in to apply all the forfeited eftates in Ireland, and the grants thereof fince the 13th of February 1688, to the ufe of the publick;

To William Bentick (lord Woodftock)	135,820
To the earl of Athlone (occafioned by the parliament of Ireland)	26,480
To the earl of Galway	36,148
To the earl of Rochford two grants of	30,512
To the lord Conningfby	5,956
To colonel Guftavus Hamilton, for his fervices in wading through the Shannon, and ftorming Athlone, at the head of the Englifh grenadiers	5,382
To fir Thomas Prendergaft for the moft valuable confideration of difcovering the affaffination plot	7,082

The report alfo obferves, that feveral of the grantees had raifed great fums of money by fale of their lands, amounting in all to £68,155; particularly the earl of Athlone (his grant being confirmed by act of parliament) has fold to the amount of £17,684, the lord Romney £30,147, and the earl of Albemarle £13000.

In thefe and moft other articles all the commiffioners agreed; but a difference arofe amongft them on account of King James's private eftate, granted to him when duke of York. This eftate three of the commiffioners, and particularly Levinge, would not allow to be forfeited, and confequently ought not to be reported. Whilft the houfe had this matter under debate, Mr. Arthur Moore, a member thereof, fent the commiffioners a letter of his own private motion, wherein he directed them to make a feparate article of the Lady Orkney's grants, becaufe that might reflect upon *fome body*, meaning the King. Mr. Montague having learned the contents of Moore's letter, and being zealous to vindicate the King's honour, which he thought ftruck at in the letter, complained of it to the houfe. Mr. Moore, being preffed to tell his author, at firft excufed himfelf, alleging that he was under a private obligation not to reveal what had paffed in private converfation, but the houfe infifting upon it, he named Lord Chancellor Methuen, who was alfo member of the houfe, who denied pofitively that he had mentioned any fuch thing. The houfe therefore refolved that the report was falfe and fcandalous, and a motion being made that the four commiffioners for Irifh forfeitures, who figned the report, had acquitted themfelves with underftanding and integrity, a warm debate arofe, and in the event it was refolved in their favour, and that fir Richard Levinge had been the author of the groundlefs and fcandalous afperfions caft upon the four commiffioners,

and

publick; and ordered a claufe to be inferted therein, for
erecting a judicature for determining claims touching the
fame. They likewife refolved, that they would not receive any petition from any perfon whatfoever, touching
the faid grants or forfeited eftates; and that they would
take into confideration the great fervices performed by the
commiffioners, appointed to inquire into the forfeited
eftates of Ireland. They alfo refolved, " that the advifing,
" procuring, and pafling thefe grants had occafioned great
" debts upon the nation, and heavy taxes upon the people,
" and highly reflected upon the King's honour; and that
" the officers and inftruments concerned in the fame had
" highly failed in the performance of their truft and
" duty." And they voted, that the faid refolution fhould
be prefented to the King in form of an addrefs; which
being done, the King anfwered, " that he was not only
" led by inclination, but thought it juftice, to reward
" thofe who had ferved well, particularly in the reduction
" of Ireland, out of the eftates forfeited to him by the
" rebellion there; that the long war occafioned great
" taxes, and had left the nation in debt, and that the
" taking juft and effectual ways for leffening that debt,
" and fupporting the publick credit, was what, in his

And prefented
to the King in
form of an
addrefs, and
the King's
anfwer there-
to.

and he was committed to the tower; however, the grant to the countefs of Orkney
was placed at the end of the report under thefe terms, viz. " a grant under the great
feal of England, dated May 30th 1695, paffed to Mrs. Elizabeth Villiers, now
countefs of Orkney, of all the private eftates of the late King James (except a
fmall part in grant to the lord Athlone) containing 95,649 acres, worth yearly
£15,995 18s. value £337,943; out of which is payable £2000 a year to Lady
Sufanna Belafyfe for her life, and £1000 a year to Mrs. Godfrey for her life; and
almoft all the old leafes determine in May 1701, when the eftates will anfwer the
values above mentioned." This report was animadverted upon by many political
tracts, and more efpecially in one entitled *jus Regium*, or the King's right to grant
forfeitures, wherein the value of the Irifh confifcations are reduced to £500,000
and the report of thefe commiffioners much expofed.

● opinion

" opinion, would beft contribute to the honour, intereft,
" and fafety of the kingdom."

This anfwer fo provoked them, that they refolved, "that whoever advifed it had ufed his utmoft endeavours " to create a jealoufy between the King and his people." They then paffed the bill of reaffumption; and ordered the report of the commiffioners for Irifh forfeitures to be publifhed; and that the refolutions of the 18th of January and 4th of April 1690, relating to the forfeitures, the King's fpeech of the 5th of January 1690, the addrefs of the houfe of the 4th of March 1692-3, and his Majefty's anfwer thereunto, be alfo reprinted with the report. And they refolved, that the procuring or paffing exorbitant grants, by any member now or formerly of the privy council, in this or any former reign, to his ufe or benefit, was a high crime and mifdemeanor.

Re-affumption bill and refolutions paffed by the Commons.

In the reaffumption bill little regard was fhown to the purchafes made under the King's grants, and to the great improvements made by the purchafers and tenants, which were faid to have doubled the value of thofe eftates.

No regard had in the re-affumption bill to the improvements made.

However, that fome juftice might be done both to purchafers and creditors, thirteen truftees were named, in whom all the forfeitures were vefted, with authority to hear and determine all juft claims relating to thofe eftates, and to fell them to the beft purchafers; and the money to be raifed to be appropriated to pay the arrears of the army. They alfo refolved, " that no perfon fhould be a truftee " who had any office of profit, or was accountable to the " King, or was a member of parliament; and that the " truftees be chofen by balloting; which being done, " the choice fell upon Francis Annefly, James Hamilton,
" John

Truftees appointed of the forfeited eftates and their powers.

" John Baggs, John Trenchard, James Ifham, Henry Lang-
" ford, James Hooper, Sir Cyril Wyche, John Cary, Sir
" Henry Sheers, Thomas Harrifon, William Fellows, and
" Thomas Rawlins."

The re-
affumption
bill confoli-
dated with
the money
bill, and
paffed into a
law.
The contefts were very warm about paffing the bill, and in the end it was confolidated with the money bill, which was to pafs for payment of the fleet and army, and under the title of a bill, " For granting an aid to the " King by the fale of the forfeited and other eftates in " Ireland, and by a land tax in England." It was then fent up to the lords, and after feveral conferences between them, and much difference, was paffed into a law *.

Eftates for-
feited in the
rebellion in
1688.
And by this act, viz. 11 and 12 Will. III. fefs. 2. c. 2. Eng. all honours, manors, lands, tenements, rents, and reverfions, in Ireland, whereof any perfons who ftood con-victed or attainted of high treafon or rebellion in Ireland, or of other treafon committed in foreign parts, fince 13th February, 1688, or fhould be convicted or attainted before the end of Trinity term, 1701, or who ftood convicted or attainted by reafon of being found by inquifition to have died or been flain in actual rebellion fince the faid 13th of February, 1688, were feized or poffeffed or interefted in, or entitled to, or any in truft for them, on the faid 13th day of February, or at any time after; or whereof

---

* Among all the hardfhips of this bill the cafe of the Earl of Athlone was moft fingular; the Commons had been fo fenfible of his good fervices in reducing Ire-land, that the addreffed the King to give him a recompenfe fuitable thereto; the parliament of Ireland had confirmed a grant made to him of between 2 and 3000l. a year; and he had fold to thofe who thought they had purchafed under an unquef-tionable title, yet no regard was had thereto, and the eftate was thrown into the heap.

the

the late King James II, or any in truſt for him, was ſeized or intereſted in at his acceſſion to the Crown of England, are veſted and ſettled in the real poſſeſſion and ſeizin of the truſtees, and their heirs, executors, &c. according to the ſeveral * eſtates and intereſts which the ſaid perſons, &c. had therein on the ſaid 13th of February, or at any time afterwards; to the end the ſame may be ſold and diſpoſed of for the uſes mentioned in the act. And where any of the ſaid perſons were ſeized of an eſtate tail only in the ſaid honours, manors, &c. the ſame are thereby enacted to be veſted in the ſaid truſtees, and their heirs, in fee ſimple, to be ſold and diſpoſed of as aforeſaid; with a ſaving for perſons compriſed within the articles of Limerick or Galway.

margin: Or whereof King James II. was ſeized at his acceſſion. veſted in truſtees.

To be ſold for the uſes in the act.

Eſtates tail veſted in the truſtees in fee.

And all grants, demiſes, ſurrenders, releaſes, cuſtodiams, &c. or diſpoſitions, ſince the ſaid 13th of February, 1688, made or granted under the great ſeal of England and Ireland, or ſeal of the Exchequer in Ireland, or by act of parliament in Ireland, or otherwiſe, of any of the ſaid forfeited or forfeitable eſtates or intereſts, or of the eſtate of the ſaid late King James, or of any of the quit rents, crown rents, compoſition rents, or chiefries, belonging to the Crown of Ireland, are thereby declared null and void.

margin: All grants, &c. of the ſaid forfeited eſtates, &c. ſince the 13th of February, 1688, void.

* In the caſe of Ellis and Segrave, in the court of Chancery here, Mich. 1758, a queſtion aroſe, as a principal point in the caſe, as to what eſtates were veſted in the truſtees by this act. Lord Bowes was of opinion that only the eſtate or intereſt, which the perſon convicted or attainted had in the lands, was veſted in the truſtees; and on this opinion granted an iſſue. But on appeal to the Houſe of Lords of Great-Britain, they were of opinion that the lands of ſuch perſons are veſted generally; and that all perſons, having reverſions, remainders, or incumbrances, were to claim them within the time preſcribed, or to be without remedy; and that the judgment of the truſtees was to be concluſive.

And

248 Of THE **EXCHEQUER** AND

Rewards to
be given to
difcoverers of
forfeited
eftates con-
cealed.

And a power was given the truftees to reward difco-verers of any fuch forfeited eftates concealed, by giving them fuch proportion of the value, after fale thereof, as they fhould think fit.

Claims to be
made of
eftates,
charges, &c.
on the lands
vefted.

And all perfons whatfoever, bodies politick and corpo-rate, having any eftate, right, title, intereft, &c. charge or incumbrance whatfoever, in or to the lands, tenements, &c. vefted in the truftees, before the 13th of February, 1688, by reafon of any fettlement, judgment, &c. affecting the faid eftates, were thereby directed, on or before the 10th of Auguft, 1700, (which time was by 12 and 13 Will. III. c. 10. Eng. enlarged to the 25th of March, 1702,) to enter their claims and demands thereto before the truf-tees; or in default thereof, every eftate, right, title, in-tereft, &c. in or to the faid premifes, was to be void, and the eftates fo liable thereto difcharged of and from the fame. And the truftees were to hear and determine fuch claims before the 25th of March, 1701.

The truftees
to be a court
of record.

The truftees to be a court of record, and their judg-ments or decrees to be entered of record in books of parch-ment to be provided for that purpofe, and to be final, notwithftanding any difability in the claimants. And all infants, feme coverts, idiots, perfons of infane memory, or beyond the feas, corporations, and all other perfons, bodies natural and politick, their heirs and fucceffors, and their interefts were to be concluded by fuch judgment *.

And

* It was determined in the King's bench here, in the cafe of Dixon and Annefley, and the judgment affirmed, upon a writ of error in the King's bench in England, Hill, 5 Ann. that the truftees had no power to determine what lands were vefted in them :

And that the truftees, upon allowing fuch claims, for the better fecurity of fuch claimant, his heirs, executors, &c. fhould give certificates under their hands and feals, containing the fubftance of fuch claim, and the allowance thereof; (which certificate, or a copy of the entry of the decree or judgment in their books, was made evidence, in all courts, of the allowance of fuch claim) and that fuch eftate, right, title, intereft, &c. or incumbrance, fo allowed, fhould never after be called in queftion by the King, his heirs or fucceffors, or by the truftees, or any claiming under them, or any of them; fubject, neverthelefs, to the power herein after given to the faid truftees concerning the fame.

Claims may be certified, and a copy good evidence.

And, after the expiration of the time for making fuch claims, the truftees were thereby directed, before the 25th of March, 1702, to fell the eftates and interefts vefted in them, and not claimed, and the eftates and interefts claimed, as foon as the claims fhould be determined; fuch fale to be made to any perfons, bodies politick or corpo-

The lands, &c. to be fold by the truftees by publick cant.

them; for that no lands were intended to be vefted in them but fuch lands as belonged to forfeiting perfons, or to King James II. which was a matter they could not determine; and that their power to inquire which were thofe lands was only in the nature of an inquifition: and that therefore if an innocent perfon claimed an eftate of inheritance before the truftees, and his claim were difallowed, he was not precluded from trying his title at law; for that their determination as to that matter was *coram non judice.* But it feemed to be admitted that their determinations as to claims of particular eftates, charges, or incumbrances, were final and conclufive. Holt's, rep. 372, 394.

But by 6 Ann, c. 34. Eng. all perfons claiming right or title to any of thofe eftates, or any incumbrances thereon, as not being vefted in the truftees, or on any other pretence, were limited to profecute their claims in two years from the 24th of June, 1708, in any court of record, or otherwife to be barred.

rate, by cant or auction. And the power to the truftees
to fell was afterwards, by 1 Ann, c. 13. Eng. enlarged to
the 24th of June, 1703.

Such as re- And by 1 and 2 Ann, ftat. 2. c. 21. Eng. all eftates
mained unfold
the 24th June, vefted in the truftees to be fold, and which were not fold
1703, vefted before the 24th of June, 1703, or otherwife difpofed of,
in the Crown
under the ma- purfuant to the former act, were vefted in the Crown for
nagement of the ufes intended by the act aforefaid, fubject to fuch
the commif-
fioners of the orders as fhould be given by the parliament of England in
revenue. that behalf; and from that day all powers given to the
truftees were to ceafe, and the truftees were to deliver up
to the commiffioners of the revenue, by indenture to be
enrolled in the Exchequer here, all deeds, records, and
papers, in their cuftody, touching the premifes: and after
that day the faid commiffioners were to levy and collect
all the rents and profits of the faid forfeited eftates, and
pay the money arifing thereby, after all charges, into the
Exchequer, there to be kept apart from all other the
King's treafure, to be applied for the ufes aforefaid, according to the orders of the parliament of England.

But the com- But it is held that the commiffioners of the revenue
miffioners
cannot make cannot make any effectual leafe of any part of thofe for-
leafes of feited eftates which remain undifpofed of, they having
them. only a power to levy and collect the rents, &c. and that
fuch leafe muft be made by letters patent under the great
feal.

Whether Although the encouragement given to the difcoverers
they can
give a re- of concealed forfeited eftates, by the power given to the
ward to dif- truftees of allowing to the difcoverers a fourth part of the
coverers. value

value of what fhould be fo difcovered, be not continued
by the laft mentioned act, nor that power transferred to
the commiffioners of the revenue, yet it is thought that
the commiffioners may, by his Majefty's directions, make
an allowance for fuch difcoveries, it being a neceffary and
incident charge relating thereto...

CHAP.

C H A P. XXIII.

OF INFORMATIONS IN THE EXCHEQUER.

Informations
in this court
what.

AN information on behalf of the Crown, filed in this court, is a method of suit for the recovery of money or other chattels; or for obtaining satisfaction in damages for any perfonal wrong committed in the lands or other poffeffions of the Crown. And it is grounded merely on the intimation of the Attorney general, who gives the court to underftand and be informed of the matter in queftion.

The different
kinds.

The moft ufual informations are thofe of *intrufion, debt,* and *devenerunt*; which latter is the Crown's action of trover. But there is alfo a particular kind of information, ftyled *in rem*, when any goods are fuppofed to become the property of the Crown, and no man appears to claim them, or difpute the title of the King.

The procefs
to iffue
thereon.

Upon the above general kinds of information the Attorney general may have an attachment for the firft procefs if he requires it; upon which the defendant is to put in bail if it be required. But the moft ordinary courfe is

by

by *fubpæna*, and procefs of contempt; and if it be againft a lord fpiritual or temporal, or a corporation, procefs of *diftringas* is to go.

If the King be feized of lands or tenements he cannot be diffeized or ejected, but if any one enters he will be an intruder upon the King's poffeffion; and therefore if a man enters upon the King's demefnes, and takes the profits, it will be intrufion; fo if he enters upon a poffeffion caft upon the King by defcent, efcheat, &c. before entry by the King; or if a man enters upon a farmer, or committee of the King; or if the King's tenant hold over his term; or if a man oufts the King's leffee for years; all thofe are intrufions on the King, for which an information will lie. ftamf. præ. 56. b. Co. Litt. 277. a. Sav. 7. 69.

Information of intrufion where it lies.

An information of intrufion likewife is the proper remedy for the recovery of eftates forfeited to the Crown, upon attainders of high treafon, or which the Crown is entitled to by efcheat. But tho' by 33 Hen. VIII. c. 20. all lands, &c. forfeited to the Crown by an attainder of high treafon are *ipfo facto* vefted in the Crown, without any office or inquifition found, yet in fuch cafe it is neceffary, for afcertaining the certainty of the lands, to have them found by office, by which they may be put in charge; which is called an office of inftruction.

For recovery of lands forfeited or efcheated.

The King by his prerogative may enforce the defendant in informations of intrufion to plead his title fpecially; and the ancient courfe of the Exchequer has been, that if in fuch informations the defendant plead " not " guilty," he fhall lofe the poffeffion. And it is faid that the reafon of this courfe is, for that regularly the

The defendant muft plead his title fpecially.

King's

King's title appears of record, and therefore the defendant may take knowledge thereof ; and the rather for that in every information of intrusion it is specified of whose possession the lands, &c. were ; but if the defendant pleads " not guilty," the King's counsel cannot know the defendant's title to provide to answer the same, as the defendant may do the King's title. 4 inst. 116. Dyer 238. Hard. 451.

But not where the King, &c. has been out of possession for twenty years.

But now by 15 Car. I. c. 1. where the King, or those under whom he claims, or others claiming under the same title, hath or have been or shall be out of possession by the space of twenty years, and hath or have not taken the profits of any lands, &c. within that space, before any information of intrusion brought to recover the same, in every such case the defendant may plead the general issue, and retain the possession until the title be tried, and found or adjudged for the King.

And no *scire facias* shall be brought in such case

And by that statute, where such an information may aptly be brought on the King's behalf, no *scire facias* shall be brought, whereunto the subject shall be forced to a special pleading, and be deprived of the benefit of the act.

Plea must conclude with a traverse of the intrusion.

The plea of a special title in the defendant must conclude with a traverse of the intrusion laid in the information. Plowd. 548.

Replication.

If the plea alleges several facts, the King by his prerogative may in his replication traverse them all, tho' a common person ought to traverse but one. Sav. 19, 64.

If

If the plea alleges a title which avoids the poffeflion in the King, fuppofed by the information, the King need not maintain the information, but may traverfe the title alleged by the plea. Sav. 61, 64. Cr. Ja. 481.

But it is fufficient if the King by his replication traverfes fo much of the title as encounters the information, without anfwering to the whole title alleged by the defendant. As if to an information of intrufion in the moiety of a manor, the defendant fays, A. was feized of the whole, and died feized of the whole, by which there was a defcent to the defendant, it is fufficient to traverfe that he died feized of fuch a moiety. Sav. 61.

The judgment, in an information of intrufion, for the King is, " that the defendant be convicted of the intrufion, &c. and be removed from the poffeffion, and be attached to make a fine; and fometimes that the lands, &c. be taken into the King's hands, and the defendant attached &c." and upon fuch judgment, every party to the information or claiming under him fhall be removed from the poffeffion; but a ftranger to the information fhall not be debarred from his entry by fuch judgment; for it does not include any judgment that the king recover the feizin. 1 Co. 40. a. 22. a. Plowd. 561. a. Sav. 35 a. Hard. 460. *Judgment.*

It is faid in Sav. 49, that upon an information for intrufion and cutting trees, or taking other valuable things, there is judgment for damages; but the reporter adds a quære.

The King may alfo, at his election, proceed by information by Englifh bill in equity for the recovery of lands to which he is entitled; and in this cafe the bill is alfo to be *Information by Englifh bill.*

in

in the name of the Attorney general, and the proceedings are to be the fame as in other Englifh bills in this court. And it is often thought more advifable to purfue fuch method, as well for difcovèry of evidence, as to avoid the partiality of juries.

Informations in debt. The King may alfo proceed againft his debtor by way of information of debt, in the name of his Attorney general, or if his debtor die, the like remedy may be purfued againft his executors, or heir and terre-tenants. Comyn. 437. Hard. 440.

On penal ſtatutes. 3 Blacks. c. 17. This information is likewife brought for any forfeiture to the Crown, upon the breach of a penal ſtatute. And the information by the Attorney general is moſt commonly ufed to recover forfeitures occafioned by tranfgreffing thofe laws, which are enaðed for the eſtabliſhment and fupport of the revenue; others, which regard mere matters. of police and publick convenience, being ufually left to be enforced by common informers in *qui tam* informations, or aðions, which may be fued for in other courts as well as the Exchequer. But after the Attorney general has informed upon the breach of a penal law, no other information can be received. Hardr. 201.

A penalty not appropriated muſt be fued in Exch. If a penalty is inflicted by ſtatute on any offence, and there be no appropriation of it, nor any method prefcribed by which it ſhall be recovered, the penalty is to be confidered as a debt to the Crown, fuable for in the Exchequer; and no indiðment will lie for the offence. Stra. 828.

Where a penalty veſted in the Crown no information in B. R. And where a penalty is veſted in the Crown only, the court of King's bench will not grant an information; but it muſt be filed by the Attorney general. Stra. 1234.

All

On an information in debt for non-payment of duties, evidence may be given of an importation on a different day from that laid in the information; but, upon an application to the court by the defendant, they will make an order for confining the evidence to a certain time. Bunb. 223.

Information for non-payment of duties, evidence may be given of a different day.

So on an information in debt for the duties of goods imported on a day certain, evidence may be given of several importations at several times. Bunb. 262. But in this case the plaintiff had given the defendant a note of the times of the importations.

Or of several days.

In an information for not making a true report, contrary to the statute, the importation was laid to be within the port of London; upon evidence it appeared that the importation was at Cowes in the county of Southampton. It was objected for the defendant that, though the information might be brought in Middlesex, yet they ought to have alleged the importation to have been according to the fact, viz. at Cowes. And of this opinion was the Chief Baron. Bunb. 261.

Information for making a false report, must be laid where the importation was.

An information upon a statute must set forth every thing requisite to bring the offence within the act; and the words " contrary to the form of the statute" will not help it; for that is only a conclusion from the premises. Bunb. 129, 177. Hard. 217.

Information on a statute must bring the offence within the statute.

All informations, as well those brought by the Attorney general, as those brought by common informers, are to be filed in the pleas office in the Chief remembrancer's office; but where a penalty is sued for in this court by way of

Informations where to be filed.

action, and not by way of information, it is to be filed in
the pleas office in the law fide of the court, where actions
are brought between party and party.

The rules to plead.

And upon thofe informations in the Chief remembrancer's office, the fecondary is to enter three rules to
plead in four days; and when the three rules are expired,
judgment is entered by default on a certificate of no plea.

Offences against the act of cuftoms determinable by the Ex-chequer.

By ftat. 14 and 15 Car. II. c. 9. commonly called
the act of cuftoms, all offences againft that act are thereby
directed to be heard and determined by the barons of
the Exchequer. And one moiety of all fines, penalties, or
forfeitures, is thereby given to the King, and the other to
him that fhall feize or fue for the fame in the faid court.
And all profecutions under the act muft be within 12
months after the offence committed.

Thofe againft the act of ex-cife by the commiffion-ers, &c.

But by 14 and 15 Car. II. c. 8. commonly called the
excife act, all offences againft that act are to be tried
before the commiffioners of excife, or their fub-commiffioners. And as many breaches of the cuftoms are
likewife offences againft the act of excife, few informations
for penalties under the act of cuftoms have of late years
been brought on the act of cuftoms.

Offences only determinable in Exchequer.

The following offences however of the act of cuftoms,
are not included in the act of excife, and can therefore
be only profecuted in this court, viz.

Receiving goods on board before declaration, or failing be-fore out-voice.

No mafter, &c. fhall receive on board any goods to be
exported, before he fhall have declared to the cuftomer,
&c. his intention to lade, and the port he is bound to;
                                                        nor

nor ſhall fail before he ſhall have outvoiced upon oath; under the penalty of £100.

No maſter, &c. ſhall break bulk until he ſhall invoice upon oath, and enter into bond that he ſhall not fail without being cleared and diſcharged by the collector, &c. under the penalty of £100.

Breaking bulk before invoicing and give bond, &c.

If any perſon ſhall refuſe to permit the collector, &c. to ſecure or take out of any veſſel any fine goods of ſmall bulk, to be put into the warehouſes of the cuſtomhouſe, till the duty be paid; or to unlade and ſecure all goods which ſhall not be unladed or diſcharged within twenty eight days after the arrival of the veſſel, he ſhall forfeit £100.

Refuſing to permit fine goods, or goods not unladen within 28 days to be brought on ſhore.

If after the clearing of any ſhip or diſcharging the officers from on board, there ſhall be found on board any goods which have been concealed from the officers, and for which the cuſtom has not been paid, the maſter ſhall forfeit £100.

Having concealed goods aboard after clearance.

L l 2 CHAP.

CHAP. XXIV.

OF INFORMATIONS IN THIS COURT ON GOODS SEIZED, OTHERWISE STYLED IN REM.

Seizure of de-
relict goods
for the
Crown.
Gilb. treat.
Exc. 180, &c.

AND firſt it is to be obſerved, that this proceeding of ſeizing goods and merchandizes for the non-pay-, ment of cuſtoms, and the like, is termed in the law a proſecution *in rem*. For the better underſtanding of which, we are to conſider that, where there was no property in lands or goods, they belonged to the Crown; and hence, if a man died without heir, and there was no tenure of his lands from any particular lord, the eſcheator ſeized them for the Crown. So all wrecks, waifs, and eſtrays were ſeized by the ſheriff for the Crown ; and in thoſe caſes, on ſuch ſeizures, they uſed to make proclamation; and if, upon the ſecond proclamation, no body came in to claim the lands or goods, they were preſumed to be derelict.

So that upon every ſeizure they were wont to file in-formations in the courts of record, and then to make the firſt proclamation, in order to condemn ſuch lands or goods to the King's uſe. And then there iſſued a com-miſſion of appraiſement, in order that the ſame might be valued, and that the ſheriff might anſwer the value thereof

thereof to the King's ufe; and upon the return of the commiffion of appraifement there was a fecond proclamation made; and then, if no body put in his claim, they were prefumed to be derelict, and forfeited to the Crown. But in the cafe of eftrays there was an abufe, by the fheriff's taking up horfes and fheep, and getting them appraifed and proclaimed, and forfeited to the Crown as derelict; and therefore a year and day was given to the owner to claim before fuch prefumption took place.

When they conftructed penal laws by way of forfei- *Forfeitures* tures, the forfeiture was appointed *in rem*, and likewife *under penal* a penalty was laid upon the perfon tranfgreffing the law; *therefrom.* and hence it was that, upon feizures, fuch goods were often derelict, becaufe the owners would not come in to claim them, left they fhould be fubject to a perfonal information; and therefore the two informations were entered; and upon the firft proclamation a writ of appraifement went out, that the officer or perfon that feized might be anfwerable for the King's part, as the claim was always entered upon the fecond proclamation.

But the proceedings in the court of Exchequer on goods *Proceedings* feized, &c. at this day, are thus; *in the Exchequer on goods feized, &c.*

When the commiffioners of his Majefty's revenue have *By informa-* directed the profecution, the folicitor of the revenue is to *tion.* file an information in the office of pleas in the chief remembrancer's office, in the name of fome fictitious perfon; but in thefe cafes, no procefs whatfoever is to iffue, either before or after the information is filed, as the feizure is deemed to be fufficient notice to the proprietor and every perfon concerned.

And

Rules to plead.

And upon thefe informations the fecondary is to enter rules to plead, as upon informations *qui tam*, *&c.* upon penal ftatutes; and the proceedings to judgment, are the fame, for want of a plea.

Proclamations.

And when the information is fo filed, the folicitor of the revenue may caufe the ufual proclamations to be made, which are thus, viz.

" If any perfons will claim property to, or fhow caufe why the fhip or veffel called ——— with her furniture, *&c.* lately feized at *&c.* being imported contrary to the ftatute, fhould not be forfeited, let them come forth, and they fhall be heard."

And on the fecond or third day afterwards, inclufive, the like proclamation is to be made; and in the fame time afterwards a third; and thefe proclamations are made in the Exchequer, fitting the court, and entered in the rule book; and the firft of them is generally made immediately after the information is filed, and before any rule to plead is entered thereon; tho it is fometimes otherwife.

Judgment for want of a plea and fale of the goods and writ of delivery.

And immediately after the third proclamation is made, and on the fame day, judgment being firft entered upon the information for want of a plea, a motion may be made by the counfel to the commiffioners for a day to be appointed for the fale of the feizure, which the court will order. And on that day the counfel to the commiffioners is to move on the faid order, that the feizure may be fold purfuant thereto, which the court will alfo order; and then the feizure is to be fet up to cant, and the higheft higheſt * bidder is declared the purchaſer, and thereupon an order is entered of courſe for a writ of delivery to iſſue for the delivery of the goods purchaſed to the purchaſer, on his paying the money to the Chief remembrancer; but generally the counſel to the commiſſioners moves the court, at the ſame time, that the Chief remembrancer may pay the money to the ſolicitor of the revenue, which the court will alſo order.

And note, in the general, where there is no claim, and eſpecially if the goods be of a periſhable nature, the ſolicitor of the revenue moves for a † writ of appraiſement, which is granted of courſe.

Writ of appraiſement for the petitioner, in what caſes.

If

---

* In Bunb. 77. it is ſaid, that if there be a condemnation without a trial, the bidder muſt ſtand to all hazards; but if after trial the bidder ſuffers by delay, the the court often diſcharges the bidder. But the reporter adds a quære. And it is likewiſe there ſaid that the court had ſome doubt what execution to order againſt a bidder not having paid his bidding, the proceſs of the pipe being, that which ſhould regularly iſſue upon an informatio nof the ſeizure; but that that being long and tedious, they ordered a *fieri facias*; as is uſual in the caſe of a perſonal information. And, in a note there, two caſes are cited, where, in ſuch caſes, the court upon affidavit iſſued attachments againſt the bidder.

† In Bunb. 30. it is ſaid, that after a ſeizure of goods, the regular ſteps are to file an information, and then take out a writ of appraiſement, upon the return of which the defendant is to enter his claim, and then may move for his writ of delivery. If the proſecutor delays filing an information, or does not ſue out a writ of appraiſement, the defendant, upon entering his claim in the book in the office, may move for a writ of delivery.

And in Bunb. 59. it is held that writs of appraiſement are a neceſſary part of the information upon a ſeizure, by the courſe of the court; beſides the act of tonnage, and poundage directs a moiety of the rates to be anſwered to the King, which ſhews there is a neceſſity for a valuation.

And

Claim of
goods feized
and form
thereof.
If any perfon would claim the goods, he may do fo
at any time before the rule for judgment for want of a
plea is made abfolute; and the claim is to be entered in
the office of pleas in the Chief remembrancer's office in
the appearance book, thus; " A. B. mafter of the fhip
or veffel called———&c. this day appeared by E. M.
his attorney, and claims the property of, &c. at the fuit
of, &c. who as well, &c."

Appearance
to be entered
with it.
So that at the fame time the claim is entered, an appearance is to be alfo entered for the defendant by his
attorney.

Rule, 24th
April, 1716,
recognizances
in what cafes
upon claims.
By a rule made in the office of pleas, in the Chief
remembrancer's office, the 24th of April, 1716, it was
ordered, that upon all informations to be exhibited for
fhips, wool, or other goods thereafter feized, no perfon,
or perfons, be thenceforth admitted to claim property in
the fame, before he, or they, enter into a recognizance,
with good fecurity, to pay the appraifed value of the
fame, the penalties in the acts of parliament made in
fuch cafes, and alfo all fuch cofts and damages, as fhall

And in another cafe there, after a condemnation and fale upon a feizure, it appearing to the court that the fpecies of the goods had not been defcribed with fufficient
certainty in the writ of appraifement, the court made a rule to fhow caufe, why
the condemnation fhould not be fet afide, and why an attachment fhould not go
againft the feizers. Bunb. 89.

Where it appears to the court that the appraifement is at more than the goods
are worth, the court will order a re-appraifement; for otherwife the feizing officer
might be undone, who muft pay the King's moiety, according to the appraifement. Bunb. 49. 185.

bc

be awarded on the profecution of any information, to be brought for the fame; unlefs the party, or parties, who claim property, fhall make it appear to the court by affidavit, that fuch fhip, wool or other goods are really and truly his or their property.

And by another order alfo made there the 5th of June, 1716, it was ordered, that where any perfon, or perfons, come to claim property in fhips, wool or other woollen goods feized, or thereafter to be feized, he or they fo claiming property fhall make it appear to the court by affidavit, that before and at the time of fuch feizure, the property of the faid fhip, wool or other goods, was in him or them; and he or they are likewife to make it appear to the court by fuch affidavit, how he or they came to have the property of fuch fhip, wool or other woollen goods; otherwife no perfon, or perfons, to be admitted to claim a property in the fame.

Rule, 5th June 1716, no perfon admitted to claim unlefs he make affidavit of the property, &c.

Now, in the cafe of Forder *qui tam, &c.* againft eight hogfheads of fugar, in this court, the 25th of November, 1734, and the 18th of June, 1735, a queftion arofe upon the aforefaid two rules, whether the claimant was not to appear in court in perfon, and claim the goods; and it was debated feveral days, but no determination was made by the court; but the practice is now to enter the claim by an attorney in manner before mentioned.

The claim to be by attorney.

If the claimant would have a writ of appraifement, he may; but he is firft to apply to the folicitor of the revenue for his confent for that purpofe, for which he is to have two guineas; and then upon counfel's motion, and on producing the confent, the court will award the

Writ of appraifement how to be obtained.

Vol. I. M m writ.

writ. If it be in vacation time, the Chief Baron, or either of the other Barons in his abfence, will upon fuch application to him at his houfe make the like rule for a writ of appraifement; which rule the Chief Baron in this cafe is to fign in the book.

Proceedings in appointing the appraifers and in executing and returning the writ.

And then the claimant's attorney is to ferve the folicitor of the revenue with the names of four merchants, or other perfons of credit, fkilled in fuch affairs, as appraifers; and the folicitor is to ftrike out two of the names, and let two ftand; and then he returns the fame, with four named by him on behalf of the revenue, and of thefe four the attorney for the claimant alfo ftrikes out two; fo that two are left ftanding on each fide; and to thefe four, whofe names are left ftanding, and are to be lodged in the pleas office in the Chief remembrancer's office, the writ of appraifement is to be directed; and they are to fummon a jury, and to hold an inquiry thereon, as to the value of the fhip, goods, wares, or merchandizes, which have been feized; and this writ with an inquifition annexed to it is to be returned into the faid office. See the forms of the faid writ and inquifition in the appendix to this work.

If the folicitor of the revenue neglects to return the names of appraifers in due time, the officer of the court will ftrike names for him, according to the method practifed in the proceedings in the equity fide of this court.

                                                                And

And upon this return of the writ of appraisement, and upon a consent for that purpose from the solicitor of the revenue as aforesaid, for which he is also to have two guineas, and upon counsel's motion thereon, a writ of * delivery will be granted upon the claimant's giving sufficient security as is usual. And thereupon the claimant, after the rule is so obtained for the writ of delivery, is to enter into security by recognizance before the Lord Chief Baron if in town, if not before either of the other Barons, in double the value of the appraisement, conditioned that the claimant shall perform and fulfil the judgment of the court upon any information brought, or to be brought, against the ship or goods seized.

Writ of delivery on return of the writ of appraisement.

And this writ of delivery is to be directed to the store keeper, collector, surveyor, or other officer, in whose custody the goods seized are; who, upon receiving the said writ, is to deliver the ship or goods under seizure to the claimant or person for that purpose named in the writ. See the form of this writ in the appendix.

To whom to be directed.

If the solicitor of the revenue should on such applications, either for a writ of appraisement, or a writ of delivery, refuse his consent, then the counsel for the claimant may, on affidavit thereof, and notice given to

The proceedings in case the solicitor of the revenue should refuse his consent either to a writ of appraisement, or of delivery.

---

* In Bunb. 21. It is held that there are two reasons for granting writs of delivery, viz delay of prosecution, and that the goods are perishable; but that these writs are discretionary in the court. Bunb. 74. It was granted for gold watches, the steel work being perishable; and in Bunb. 20. It is said that no certain rule is laid down what shall be called delay; but that what was most generally agreed upon was, that where a seizure was in the vacation time, and there is no information filed in the term following, if the prosecutor could have tried it that term, this would be a delay to ground a writ of delivery upon.

the

the folicitor of the revenue, make fpecial application to the court, and they will either grant or refufe the writ as they fee caufe.

In the cafe of Forder *qui tam* againft John and James Wolfe, in this court, the 23d of November, 1734, a writ of delivery was refufed, as the evidence for the feizure depended in a great meafure on the manner of packing and making up the goods.

When the goods are fo claimed the defendant is alfo to plead to the information; and if it be an iffuable plea, as it ufually is, then the record is to be made up, and the after proceedings are, as on informations *qui tam* upon penal ftatutes, pretty much the fame with the proceedings in the common law fide of this court between party and party.

But as has been already obferved, moft of the offences under the act of cuftoms being likewife offences under the act of excife, informations on feizures in this court are very rare; they being moftly brought before the commiffioners or fub-commiffioners of excife.

The following offence under the act of cuftoms, however, feems to be cognizable only in the court of Exchequer, viz. that of fhipping native commodities coaftwife, without making a declaration to the collector, &c. of the contents, value, &c. and giving a bond conditioned to difcharge them in the realm; by which a forfeiture of fuch goods is incurred.

By

By the act of cuftoms there is a provifion, that, for the avoiding of fraudulent compofition, no action, bill, plaint, or information be exhibited or proceeded on againft any goods, wares, or merchandizes feized, until fuch feizure fhall be regiftered and entered with the regifter or officer to be apppointed for that purpofe in the port of Dublin, and certified by him to be fo entered and regiftered; and until fuch goods, wares, and merchandizes, be fecured or laid up in his Majefty's ware-houfe, at the cuftom-houfes of the refpective ports. And in cafe the commiffioners of the cuftoms fhall be diffatisfied, or apprehend any neglect or delay in any perfon or perfons to fue for or profecute in any action, bill, plaint, or information, as aforefaid, that it fhall and may be lawful to and for the faid commiffioners to appoint any other perfon or perfons, whom they fhall think fit, to profecute; which other perfon or perfons fhall be and are thereby declared to be true, proper, and lawful profecutors or feizers to all intents and purpofes whatfoever, and to whom the moiety of the faid feizures and forfeitures fhall be due and payable, and to none other; any thing in the faid act, or any other law, ftatute, ufage, or cuftom, to the contrary thereof, notwithftanding.

By rule 36, annexed to the act of cuftoms, all officers whom it may concern in their refpective places, fhall be diligent and careful to make ftay and feizure of goods, wares, and merchandizes, that fhall be brought in, or carried out, or intended to be carried out of this realm, contrary to the laws of the fame.

And

Marginal notes:

Seizures to be regiftered and fecured in the King's warehoufe before any proceeding by bill, &c.

Perfons neglecting or delaying to fue, the commiffioners may appoint a profecutor, who is declared the lawful feizer, and fhall have the moiety of the feizure.

Officers to be careful to feize any goods brought in or carried out contrary to law.

Goo's feized to be put into the warehouse, and there kept until released by sufficient warrant.

And by rule 37, all goods and merchandizes that shall be seized or staid shall, presently after such seizure or stay, be delivered into the charge of the ware-house keeper at the custom-house of the port where such stay or seizure shall be made, there to remain until sufficient warrant and discharge shall be brought for release and delivery thereof.

Officers making seizures forthwith to acquaint the commissioners therewith, and to certify the same to the register of the seizures.

And by rule 38, every officer, who shall make any seizure, shall thereupon forthwith acquaint the commissioners of the customs therewith, and likewise certify the same to the register of seizures in the port of Dublin for the time being, together with the quantity and quality of the goods so seized, the time when, the ground whereupon he seized the same, with such other circumstances as are fit to be known, for exhibiting informations in the Exchequer against the same.

No officer to compound a seizure without licence or other lawful warrant.

And by rule 39, no officer or other person shall make any composition or agreement for the seizure or forfeiture of any goods, without * licence out of the court of Exchequer, or other lawful warrant first had and obtained.

And

---

* I do not find that the taking out of these licenses hath been practised here these many years. Lord Chief Baron Gilbert, in his treatise of the court of Exchequer in England, page 186, &c. gives the following account of them:

When a suit (says he) was commenced, even between party and party, they could not compound the same without leave of the court, which was the original of all fines concerning lands and tenements; and the reason was, because the K an interest in every suit in his court, since there was an amerciament in most; much more in informations, where the King himself was party, so such an interest that the informers could not compound without leave

And by rule 40, all licenfes, compofitions, fines, recoveries, warrants, orders, and other difcharges, to be had, made, or granted for or upon the aforefaid feizures and informations, are to be entered with the regifter aforefaid, and the money or monies thereupon due and payable to the ufe of his Majefty to be paid to the collectors of the refpective ports.

Licenfes, &c. for forfeitures to be entered with the regifter, and the money due to the King to be paid to the collector.

And by rule 41, all appraifements of goods, wares, and merchandizes, feized as aforefaid, are to be fhowed and delivered to the regifter aforefaid, before they be returned

Appraifements of feizures to be delivered to the regifter for examination and entry.

court; but yet in many cafes, where penalties were great, and the offenders poor, it would have been exceedingly hard if the law had been inexorable, and the informer might not have compounded with the offender; and it would have ftill been more derogatory to the honour of the Crown if the informer had compounded, and there had been no method found out to have made a compofition for the Crown. From hence it is that there is a ftanding privy feal, by which the commiffioners of the treafury, High treafurer, Chancellor, Under treafurer, Chief Baron, Barons of the Coif, and Attorney general, or any one of them, are empowered to give a licenfe to compound; provided no fine be fet lefs than half fo much as the informer fhall or is to have for his part. In order to fee that the King's part be at leaft equal to one half of the informer's, there muft be an affidavit made by the informer of what he receives upon fuch compofition, and then they go back to the officer, and the compofition is fet, and then it is carried to be figned by the commiffioners of the treafury, Lord High treafurer, Chief Baron, Attorney general, or any two of them, who by the faid privy feal are entitled to compound the fame.

This power was abufed by offenders againft penal ftatutes; for after fuch tranfgreffions they ufed to fet up fham informers in order to get rid of the penalty, and fo compound with them for a little, and diminifh the King's part almoft to nothing: for this caufe it was that by the rules of the court the chriftian and furnames, with the addition of the parties, are to be put into the licenfe, together with the place of their abode; the licenfe is to be figned by a fworn clerk, and entered in a book before the fame is figned by a Baron.

And how long this licenfe to compound is to be in force, how the compofition fhall be recorded, the fine rated and paid, and a writ of delivery obtained thereon, and how this writ is to iffue where the fine is paid, and how where fecurity is given, fee ibid. pag. 188 to 191.

into

into the Exchequer, to be by him examined and entered.

If the goods are under-valued a new appraisement to be made.
And if the goods be too much undervalued, the faid regifter is to make ftay thereof, and to acquaint fome of the Barons of the Exchequer therewith, to the end that a review and new appraifement may be made of the goods.

Coaft bonds, for which certificates are returned, to be deli-vered quar-terly into the Exchequer.
And by rule 42, all bonds taken for fhipping goods to the coafts, for which certificates are returned, fhall be deli-vered quarterly into the Exchequer, with the certificate thereunto annexed and endorfed alfo thereupon ; and every term, after the accompt of the officers that did take them is paft, the faid bonds fhall be delivered to every perfon that fhall fue for the fame, paying the ufual fees.

All other bonds to be delivered into the Exche-quer after the breach of conditions to be put in fuit.
And by rule 43, all other bonds taken by the collectors that be expired, and all other bonds for which no certifi-cates are returned, according to their conditions, fhall be delivered likewife into the Exchequer quarterly, after the breach of fuch conditions, that procefs and execution may be had thereupon according to the due courfe of law.

CHAP.

C H A P. XXV.

OF INFORMATIONS BEFORE THE COMMISSIONERS OR SUB-COMMISSIONERS OF EXCISE.

BY 14 and 15 Car. 2. c. 8. commonly called the act of excise, an office is created in the city of Dublin, to be called by the name of the office of excise or new impost, and to be managed by commissioners not exceeding five in number, and also a surveyor; all to be appointed by the Lord Lieutenant or Chief governors of the kingdom.

Commissioners of excise created in Dublin,

And the like offices, and in them such sub-commissioners or collectors, are thereby directed to be appointed in all the counties of the kingdom, and in all other cities and places thereof, as the commissioners shall think fitting, to be approved of by the Lord Lieutenant or Chief governors of the kingdom.

and sub-commissioners in the country,

And the commissioners or collectors of excise in their respective districts, or such other persons as shall be authorized thereto, together with such sub-commissioners or collectors, are thereby authorized to * hear and determine all offences and breaches of any clause in said act, other than such as are otherwise thereby appointed; and are,

to hear and determine offences against the act of excise.

---

* When a day of trial is appointed by the commissioners or sub-commissioners, the constant practice has been to give the claimant eight days notice thereof, exclusive of the day on which the summons or notice of trial is served, and inclusive of the day of trial, as on trials by *nisi prius* in the four courts in the county of the city, or county of Dublin.

upon notice or information, to proceed to examination of the matter in fact, by fummoning parties and * witneffes to appear before them, and examining witneffes upon oath in the prefence of the party accufed, if he appear; and in cafe he fhall neglect to appear, they are authorized to proceed as if he were prefent: and upon proof of the fact, by the confeffion of the party, or oath of one credible witnefs, they are authorized to give judgment, and iffue a warrant for levying any forfeiture, fine, or penalty, inflicted by the act, by diftrefs of the party's goods, or in default of fufficient diftrefs, to commit the party to prifon until he pay it.

Informations for penalties to be within fix months.
But all informations for any penalty incurred by this act are to be made within fix months after the offence fhall be committed.

An appeal given to the Lord Lieute-nant, or com-miffioners of appeals.
And it is thereby provided, that if any perfons fhall judge themfelves aggrieved with any proceedings had by the commiffioners, &c. it fhall be lawful for every fuch perfon to make his † appeal to the Lord Lieutenant, &c. or fuch as he fhall appoint by commiffion under the great feal; who are empowered to fend for parties and wit-neffes, and all writings, &c. and to examine upon oath and determine all appeals, and confirm or reverfe all judg-ments given by the commiffioners, &c. and to difcharge any

---

* By 33 Geo II. c. 10. witneffes may be fummoned to appear before them tho' refiding in another diftrict; provided that no fuch fummons fhall iffue until it fhall appear by affidavit before one of the commiffioners, or fub-commiffioners, that the perfon fummoned is a material witnefs.

† By 33 Geo. II. c. 10. fuch appeal muft be brought within two calendar months after the fentence given.

perfon

perfon committed by the commiffioners, &c. and to mitigate all fines, penalties, and forfeitures, impofed by them;
provided that in the mitigating fuch fine, &c. the informer
may be duly encouraged for his pains and difcovery, according to the nature of the fraud difcovered.

And it is thereby enacted, that if any goods feized fhall *Goods feiz'd and not claimed in 21 days, to be fold.*
not be claimed or cleared within twenty-one days, the
commiffioners or fub-commiffioners, &c. appointing a general day of fale, and giving publick notice thereof, fhall
caufe the goods to be appraifed by two fworn officers, or
others, and afterwards fell them by the candle to the
higheft bidder.

And of all feizures, fines, forfeitures, and penalties, *One moiety of the fines, &c. to the King, the other to the informer.*
mentioned in this act, the neceffary charges for recovery
thereof being firft deducted, one moiety is to be to the ufe
of the King, and the other to the perfon who fhall feize,
or give any information of and prove any breach of any
claufe therein.

When a feizure is made of any goods upon the act of *The proceedings upon a feizure under this act.*
excife, if it be in Dublin, the feizing officer is to bring
them to the ftores at the cuftom-houfe, and to make a
return thereof in writing to the commiffioners, and alfo
to the regifter of the feizures; and if any petition be preferred to the commiffioners, it is referred to the feizing
officer, and on his report it either is or is not ordered to
ftand a feizure. Then the feizing note is fent by the regifter of the feizures to the clerk of the informations in
the faid port, who, if it be a general or common cafe,
either as to goods exported or imported, is to prepare an
information according to the general forms; which fee
hereafter in the appendix.

N n 2   If

If it be a cafe attended with any fpecial circumflances,
it is to be brought to the folicitor or to the commiffioners, who is thereupon to prepare inftruclions for the
counfel for the commiffioners to draw a proper information thereon. For forms, or precedents, fee the appendix.

Seizures not
claimed by
the proprietor
within twenty
one days for-
feited and
may be fold.

If the owner or proprietor of the goods do not claim
them within twenty one days, the courfe is to collect all
the feizures of the fame kind unclaimed, and to infert
them in one information, and to enter judgment thereon
for want of a claim, and then to fell at the next general
fale to be appointed by the commiffioners, or fub-commiffioners.   But this method of inferting many feizures
in one information feems liable to great objections.

If a penalty
fued for there
muft be a fe-
parate infor-
mation.

And if on any of the feizures a penalty is recoverable
and fued for, a feparate information muft be entered
for thefe goods, and a judgment of condemnation had
thereon, to be ready to be read in evidence on the trial
for the penalty.

Proceedings
before fub-
commiffio-
ners the fame
as before the
commiffio-
ners.

If the feizure be in any other port than the port of
Dublin, the informations and proceedings are to be the
fame before the fub-commiffioners as before the commiffioners of excife.

The fame
proceedings
in all cafes
relating to
the inland
excife.

And in the cafes of brewers, vintners, ale-houfe keepers,
diftillers, or retailers of ftrong waters, and all cafes
whatfoever relating to the inland excife, the like proceedings are to be againft offenders, by information, fummons, &c.

By

By the 33 Geo. II. c. 10. it is recited that claims had been frequently made of goods feized by perfons who never appeared after making fuch claims, but left the kingdom or the diftrict where the feizure was made, and were not to be found, fo as to be ferved with a notice or fummons for trial as the law directs; by reafon whereof feveral parcels of goods had remained under feizure for many years, and until they perifhed, on account of not being duly condemned, to the prejudice of his Majefty and the informer; for remedy whereof, it is thereby enacted, that in all cafes where a feizure fhall be made of any goods, &c. and a claim fhall be tendered by the owner, or proprietor thereof, or by any perfon deputed to make fuch claim, that the perfon tendering fuch claim fhall at the foot thereof mention fome particular houfe within the diftrict where the goods are feized, where notices or fummonfes fhall be left or ferved; and in default thereof that the claim fhall not be deemed legal or received, but it fhall be lawful to proceed to the condemnation of fuch goods, in fuch manner as by law may now be done for want of a claim; and that all notices or fummonfes ferved or left for fuch claimants, with any perfon above the age of fixteen years refiding at fuch houfe as fhall be fo mentioned or exprefled at the foot of faid claims, or pofted on the door eight days before the time appointed for determining the claim, if no perfon refides therein, fhall be as valid and effectual as if the perfons making fuch claim were perfonally ferved with fuch notices or fummonfes.

When a claim is made, mention is to be made at the foot of the claim of fome houfe, within the diftrict, where notice fhall be ferved.

And

Difputes con-
cerning rights
of feizure.

And by the faid act of 33 Geo, II. c. 10. reciting that, where two or more perfons have been concerned as informers or difcoverers, feveral difputes have arifen between the parties pretending to be the real informer, and difcoverer, to the great detriment of his Majefty's revenue, and difcouragement of fuch informers; and that a juft diftribution of the rewards given to fuch informers would be a great encouragement to the trade of

Commiffio-
ners and fub-
commiffioners
in their dif-
tricts to de-
termine the
right to fei-
zure, &c.

this kingdom, it was enacted, that in every cafe, where two or more perfons fhall claim any right to any reward, for or on account of any feizure, penalty, or forfeiture, they may be entitled to, the commiffioners, or fub-commiffioners in their feveral diftricts, who fhall hear and determine fuch feizures, fhould hear the feveral claims and demands of fuch perfons, as may think themfelves entitled to any reward, for, or upon account of any information, or difcovery, and give or diftribute the fame in fuch manner, or proportions, as they fhould order and direct; which order, or fentence, fhould be final and conclufive to the faid parties.

Sub-commif-
fioners to take
an oath be-
fore hearing
a caufe (if
required) that
they are not
interefted in
the feizure.

And it is by the faid act enacted, that the fub-commiffioners, collectors of excife, and other perfons that may be authorized, and appointed to hear and determine the matter of complaint mentioned in fuch information, and every of them, fhall, if thereto required by the party or parties againft whom fuch information is made, take an oath that he is not interefted, directly or indirectly, in the matter or complaint then depending before them, and that he is not to gain or lofe thereby on any account whatfoever ;

whatfoever; which oath the clerk or regifter of the feizures and forfeitures in the particular diftrict is thereby authorized and required to adminifter; and if fuch fub-commiffioner, collector of excife, or any other perfon, to be fo appointed fhall refufe to take the faid oaths, fuch fub-commiffioner, collector, &c. fhall be difqualified, and rendered incapable to hear, determine, or give judgment upon the matter then depending before them, and contained in fuch information, and all proceedings to be had before them after fuch refufal fhall be null and void.

And by the faid act, reciting that the profecutions before the commiffioners of appeals, tho' carried on in a fummary way, purfuant to the laws in force in this kingdom for that purpofe, had been artfully delayed by perfons profecuting the faid appeals, upon account of fome informality, or defect of form in the proceedings, to the great difcouragement of the profecutors, or informers, it is enacted, that no judgment, or fentence of the commiffioners, or fub-commiffioners of excife, fhall be reverfed for any informality, imperfection, or defect in form, either in the information, proceedings, or judgment brought before or given by the faid commiffioners, or fub-commiffioners refpectively.

No judgment of the commiffioners or fub-commiffioners of excife to be reverfed for informality, &c.

And whereas it often happens that the claimants of goods feized by the officers of his Majefty's revenue, on condemnation thereof by the chief commiffioners or fub-commiffioners in their refpective diftricts, enter appeals againft fuch judgments of condemnation, in order to delay the fale of fuch goods fo condemned as aforefaid, that they may thereby perifh, and his Majefty and the feizing officer lofe the benefit of the faid feizure, for

Perifhable goods feized to be fold twenty one days after the condemnation notwithftanding any appeal.

remedy

remedy thereof, it is by the faid act enacted, that all
perifhable goods, and commodities, which fhall be feized
by any of the officers of his Majefty's revenue, or other
perfon, or perfons, and condemned as aforefaid, fhall and
may be fold as the law directs, at any time after the expiration of twenty one days after the condemnation
thereof, by order of the Chief commiffioners of his
Majefty's excife, notwithftanding any appeal brought, or
to be brought, from the faid fentence of condemnation,
fix days notice being previoufly given in manner herein
before mentioned to the claimant, or left for him at his,
or her ufual place of refidence, and an affidavit being

And the pro-
duce to be
accounted for
and paid to
fuch perfons
as are legally
entitled to the
fame.

thereof made ; and the produce arifing by or from fuch
fale to be accounted for and paid to fuch perfon and
perfons as fhall be by law entitled thereto, in ten days
after the time given by law for appealing fhall be elapfed,
or in cafe of any appeal in ten days after the fentence
of condemnation fhall be affirmed, or the appeal difmiffed ; and that in cafe of a reverfal of fuch fentence of
condemnation, the produce arifing by or from fuch fale
fhall, in ten days after fuch reverfal, be accounted for
and paid to the owner or owners refpectively of the
goods fo feized and fold, in full fatisfaction for the goods
fo feized.

All forfeit-
ures and pe-
nalties inflict-
ed by this act
to be fued for
as prefcribed
by the act of
excife.

And it is by the faid act alfo enacted, that all the forfeitures and penalties thereby inflicted (other than fuch
as are otherwife thereby appointed) fhall and may be fued
for and recovered, levied and applied, in fuch manner and
form, and by fuch ways and methods, as are prefcribed
and appointed in and by the act of excife.

By

By ſtatute 1 Geo. III. c. 7. the aforeſaid act of the 33 1 Geo. II. and all and every the clauſe and clauſes therein contained (except ſuch clauſe or clauſes as are thereby altered or repealed) are continued for the ſpace of two years from the 24th day of **June** 1762 and to the end of the then next ſeſſion of parliament.

Act of 34 Geo. II continued.

And by the ſaid act it is enacted, that it ſhall and may be lawful to and for the commiſſioners of appeal, under their hands and ſeals, from time to time, to authorize and empower ſuch perſon or perſons as they ſhall think fit, in the ſeveral counties of this kingdom, to be commiſſioners to take and receive affidavits, concerning any cauſe depending, or other proceedings in cauſes of appeal, before the commiſſioners of appeal; and all affidavits taken as aforeſaid, ſhall be of the ſame force, as affidavits taken before the ſaid commiſſioners of appeal are, or may be; and for the ſwearing and taking of every ſuch affidavit, the perſon ſo empowered, or taking the ſame, ſhall receive a fee of one ſhilling and ſix pence, and no more.

Commiſſioners of appeal authorized to grant commiſſions for taking affidavits in the ſeveral counties in the kingdom.

And no affidavit taken by any commiſſioner, authorized as aforeſaid, ſhall be read or made uſe of before the commiſſioners of appeal, unleſs the commiſſioner or perſon that takes the ſame, mention in the caption thereof the day of the month when, and alſo the place and county where, the ſame ſhall be ſworn, and that he knows the deponent, or has been credibly informed that he is the real perſon mentioned and deſcribed in ſuch affidavit.

Directions concerning the caption of ſuch affidavits.

And whereas by the aforeſaid act of exciſe, or new impoſt, all goods and merchandizes ſeized for being run, or intended to be run, were to be brought to the office of excife next adjoining to the place where fuch goods were fo feized, there to be detained and kept, until the fame fhould be condemned or difcharged, in manner as by the faid act of excife is provided, which had, in many cafes, been attended with inconveniencies and damage to the owners of fuch goods and merchandizes, by lofing their market, before a trial could be had thereon; and many difadvantages had alfo arifen, by the detention of fhips or veffels laden with fuch goods and commodities, thereby preventing them from proceeding on their intended

Owners of goods feized, and mafters of veffels feized, for breach of the excife laws, may apply for a writ of appraifement, as in cafes to be heard and determined in the Exchequer.

voyages; for remedy thereof it is enacted, that it fhall and may be lawful to and for the owners of any goods, feized for being run, or intended to be run, and to and for the mafter or commander of any fhip or veffel feized for the breach of any of the laws of excife, to apply (as by law may now be done, in cafes to be heard and determined in the court of Exchequer) for a writ of appraifement, to value and appraife fuch goods and merchandizes, and fhip or veffel fo feized, on which fuch proceedings fhall and may be had, as have been ufual in cafes where by law writs of appraifement have iffued; and that on return of

And on return thereof, a re- cognizance to be entered in- to in the court of Exchequer.

the appraifement, or value of fuch goods and commodities, and of fuch fhips and veffels, the party or parties applying for fuch writ of appraifement, together with two fufficient fureties, fhall enter into a recognizance to his Majefty in double the value of fuch appraifement, before the Chancellor, or one of the Barons of the court of Exche- quer, or before fuch perfon or perfons as they, or any of them, fhall appoint by commiffion to be iffued out of the faid court of Exchequer, conditioned to pay fuch appraifed value, and all other penalties and forfeitures attending fuch such seizure, in case the same shall be condemned; and that thereupon the chancellor, or any of the barons of the said court of Exchequer, shall award a writ of Delivery in the usual manner for such goods and merchandizes, and the ship or vessel so seized as aforesaid.

Writ of De-
livery.

Provided always, that upon the acquittal of such goods and ships or vessels from such seizure as aforesaid, by the chief commissioners of the revenue, or their sub-commissioners, in their several and respective districts, or by the commissioners of appeal (in case an appeal shall be brought) and due proof made thereof before the said Chancellor, or any of the Barons of the said court of Exchequer, and notice given to his Majesty's Attorney general for the time being, that then the said Chancellor, or any of the Barons of the said court of Exchequer, shall and may order the said recognizance to be vacated; and the same shall afterwards be null and void to all intents and purposes whatsoever.

On acquittal
the recogni-
zance to be
vacated.

And by the said act it is also enacted, that all the forfeitures and penalties inflicted thereby shall and may be sued for and recovered, levied, and applied in such manner and form, and by such ways and methods, as are prescribed and appointed in and by the act of excise.

Penalties and
forfeitures in-
flicted by this
act to be sued
for, as directed
by the act of
excise.

And the aforesaid statute, as to the several matters herein before-mentioned, is to continue and be in force for two years, from the 24th day of June, 1762, and from thence to the end of the then next session of parliament.

And

The said act and the stat. 33 Geo. II. further continued.

And by the statute of 3 Geo. III. c. 21. the aforesaid act, as also the statute 33 Geo. II. c. 40. are further continued for two years from the 24th day of June, 1764, and from thence to the end of the then next session of parliament.

One commissioner of excise empowered to hear and determine complaints for selling spirits without licence.

And by the said act it is enacted, that it shall and may be lawful to and for any one or more of the chief commissioners of excise to hear and determine all complaints, and to levy all forfeitures that shall be made or incurred by or against any person or persons selling * spirits without licence, in the same manner, and as effectually, to all intents and purposes, as any three of the said chief commissioners were then empowered to do, with such remedy of appeal as is therein mentioned.

The three last acts continued.

And by the stat. of 5 Geo. III. c. 16. the said three last acts, and all and every the clauses therein respectively contained, (except such parts thereof as are altered or amended by this act) are continued for two years from the 24th of June, 1766, and from thence to the end of the then next session of parliament.

The four last acts continued.

And by the stat. of 7 Geo. III. c. 27. the said four acts are continued for two years from the 24th day of June, 1768, and from thence to the end of the then next session of parliament.

* Or, by 13 Geo. III. c. 8. wine, cyder, beer, or ale, by retail without licence.

By

By ſtat. 11 Geo. III. c. 7. it is enacted, that no writ of replevin, writ of *deliverance*, or writ of *re-caption*, ſhall, at any time hereafter, without leave firſt obtained for that purpoſe from his Majeſty's court of Exchequer, be executed for any goods or chattels ſeized by any officer of exciſe, for being run, or intended to be run, without payment of duties due and chargeable thereupon, to his Majeſty; or for goods and commodities detained to anſwer the payment of duties, due and chargeable thereupon, to his Majeſty, unleſs ſuch goods and chattels ſhall be firſt acquitted by due courſe of law.

And by ſaid act the ſaid five former acts (except ſuch parts thereof as are thereby altered, repealed, or amended,) are continued for the ſpace of two years from the 24th day of June, 1772, and from thence to the end of the then next ſeſſion of parliament.

And by ſtat. 13 Geo. III. c. 7. the ſaid ſeveral acts (except ſuch parts thereof as are altered, repealed, or amended thereby) are continued for the ſpace of two years from the 24th day of June, 1774, and from thence to the end of the then next ſeſſion of parliament *.

* One can ſcarce avoid lamenting that any neceſſity ſhould ever have happened to cauſe the inſtitution of a judicature, which ſo much ſeems to claſh with the ſpirit and genius of the Britiſh conſtitution, as that which is created by the exciſe laws; an inſtitution by which that bulwark of Britiſh liberty, a trial by jury, is partly ſubverted; and the determination of property, ſometimes to a great amount, transferred from the eſtabliſhed courts, to perſons who in the general, cannot either from their courſe of education or experience, be ſuppoſed to be acquainted with the modes of legal reaſoning, or the proceedings of juſtice.
Wherefore, the commiſſioners and ſubcommiſſioners who are the judges appointed by thoſe laws, are ever to bear it in mind, that although thoſe revenues are to be duly collected, and although none of the rights of the Crown are to be remitted, yet that the ſcale of juſtice is to be ho'den with an even hand, between the Crown and the ſubject, and that the rights of both are to be determined according to the rules of law and juſtice, for the ſafety and advantage of both.

Another

Another circumstance attending the trials upon these laws, apparently re-pugnant to the ordinary course of proceedings in the superior courts of justice, is that of admitting the testimony of the informer, who is to receive a moiety of the penalty or forfeiture, as he is swearing under one of the strongest temptations to perjury; and yet, were it not so, and that the officers were not to seize on in-formation, or detection of frauds until they could procure persons to attend them for evidence, there would be but very few convictions on these laws. Wherefore also the judges upon these trials, when the party so interested is the only evidence to be had, are to act with all the caution their prudence and discretion can suggest to them, and especially as the advantage is not reciprocal, the testimony of a trader, even of the fairest repute, being never to be admitted where he is himself the defendant upon any trial on these laws; nor in truth ought it to be, as the same necessity cannot be urged on the one side as on the other.

Not that whilst this most important office shall be exercised by men of liberal and generous minds, a contrary conduct can be apprehended; yet should it happen otherwise, and that any of these persons indiscreetly warmed by the zeal of office, or influenced by any other as improper motives, should consider themselves not merely ministers of justice, but servants of the Crown, and as such in duty bound to multiply forfeitures; or should neglect the modes of legal proceedings, it is easy to conceive what injury might be the consequence of their decisions, and how far these revenues (which are in fact granted for the publick) might then be-come what never was the intention, an engine of oppression to the fair trader, and be nearly as great an injury to those revenues as suffering transgressors to escape with impunity.

It is true, an appeal is given to other commissioners who are generally of the profession of the law; but when the delay, the heavy expence, which cannot but attend such a step, with the frequent consequential losses, are considered, as also, that the appellant has the whole weight of the revenue to contend with, and that no cost is to be paid (which is the case on both sides in these suits let the litigation be ever so groundless) it must be confessed, that the contest may be very unequal and the remedy not adequate. Besides, although actions for damages have been maintained, where the sentence below has been against the informer or seizer, yet it is otherwise on the reversal of such sentence when in his favour; it having been always deemed a reasonable foundation for the prosecution on these laws. So deter-mined in the case of Reynolds against Kennedy, 1 st Wilson, 232, B. R. which see, as also, the cases therein cited with the reasons at large.

It is also to be wished on the other hand, that in cases where the constitutional mode of proceeding hath been preserved, juries would seriously consider, that the prevention, restraint, or punishment of frauds or impositions in the payment of the duties, is not only a benefit to the publick by the augmentation of the revenue, but likewise to the fair trader, who, should such frauds be permitted, or the punishment of them eluded, could no longer subsist; and that informers and seizing officers, who are absolutely necessary for these purposes, are only blamable for what they illegally and wantonly or wickedly do in the execution of their offices; and when that is the case, every unprejudiced person cannot but admit, that the injured party is most justly entitled to an adequate recompense in damages upon any action or suit for the purpose.

CHAP.

C H A P. XXVI.

Of DEBTS due to the KING, and his REMEDIES for RECOVERY of THEM.

THIS being a fubject not very diftinctly treated of in law books, and the ftat. of the * 33 Hen. VIII. c. 39. in England, which is not in force in this kingdom, having made feveral alterations in the common law there with refpect to this matter, it will be neceffary to confider it with great caution, advancing nothing but what is fupported, or feems to be inferred from the beft authorities, and leaving a full difcuffion of the fubject to abler hands. And for the fake of method and perfpicuity it will be neceffary to arrange what feems moft material on this head under the following particulars, viz.

---

* 'A doubt has been fometimes entertained whether this act be not in force in this kingdom, fo far as it relates to the prerogatives of the Crown, by virtue of an act made here in the fame year of that King's reign, by which the King of England, his heirs and fucceffors, are to have the ftyle, title, &c. of Kings of Ireland, with all pre-eminences, *prerogatives*, dignities, &c. to the eftate and majefty of a King imperial appertaining. But this notion is deftitute of any foundation ; the latter ftatute plainly being intended to change the eftate and dignity of Lord of Ireland to that of King, without enlarging in any refpect his legal prerogatives ; much lefs thofe which he derived under a ftatute not then exifting ; the feffion of parliament in which the Irifh act was made having commenced the 13th of June, 1541, and ended the 20th of July following; whereas the feffion of parliament in which the Englifh act was made did not commence until 16th of January following. And furely whoever confiders the fpirit of tyranny and inconfiftency which marks the Englifh laws of that reign, would not be very ftrenuous to contend for their exiftence in this kingdom by any ftrained inference. The one in queftion particularly contains feveral claufes which feem very obfcure and almoft unintelligible. *(margin: 33 Hen. VIII. fefs. 1. c. 1.)*

First,

First, of the King's debtors and his remedies against them.

Secondly, of the King's precedence with regard to execution.

Thirdly, of the King's prerogative with regard to the debtor of his debtors.

And as to the first particular, viz. of the King's debtors and his remedies against them, it appears by the ancient usage of the court of Exchequer, that from the earliest ages the Crown claimed and exercised several very great prerogatives with regard to the recovery of its debts.

King could protect his debtor.

The King could grant a writ of protection to his debtor, that he should not be sued or attached until he paid the King's debt. But this was productive of great inconvenience ; for, to delay other creditors, the King's debts were the more slowly paid. For remedy whereof, by 25 Ed. III. c. 19. Eng. it was enacted, that other creditors might have their actions against the King's debtors, and proceed to judgment; but not to execution, unless such creditor should take upon him to pay the King's debt, and then he might have execution for both debts. 1 Inst. 131 b. F. N. B. 28. Dyer 328. But such protection would not lie after a suit commenced. Hard. 26. Nor could the debtor avail himself of this privilege without having the writ of protection. Cr. Car. 389. And see Hob. 115. where it is said that the restraint of the subject, as to proceeding to execution, imposed by the statute, relates to executions on lands and goods, and not of body.

And

And the King's debtor could not make a will to difpofe of his chattels to the King's prejudice; nor could his executors have adminiftration of his chattels without permiffion from the King, or from the juſticier, or the barons of the Exchequer; which they obtained upon giving fecurity to pay the King's debt: and if the debt claimed by the King were a doubtful one, the King would fometimes command the executors to retain in their hands fo much as the fum amounted to, till the matter was difcuſſed in the Exchequer. Madox 663, 664, 665. 2 Ro. ab. 158 H.

His debtor could not difpofe of his chattels by will to the King's prejudice.

If one died indebted to the King, and it were doubtful whether the chattels of the deceafed would amount to fatisfy the debts due to the King and to other perfons, it was ufual for the King to feize into his hands the chattels of the deceafed, in order to have a fatisfaction of his debt, before any other creditor of the deceafed was paid, or the chattels were eloigned, or applied to any other ufe. But when he fo feized them, he allowed a competent part for the decent funeral of the deceafed. Madox 665. See magna charta, c. 18. 2 Inft. 32.

King could feize the chattels of his debtor deceafed.

At common law, if a common perfon obtained a judgment for debt or damages, he could not have the debtor's body or his lands during his life in execution; but the body, * lands, and goods of the King's debtor were liable. 3 Co. 12.

King could have execution againſt body, lands, and goods.

* But in Palm. 167 it is doubted whether it is not merely by the cuſtom of the court of Exchequer that lands can be extended for the King's debt, and not upon the judgments of any other courts, except the debt be eſtreated into the Exchequer.

P p

The

Against heirs, terre-tenants, executors,&c.

The King might levy his debt not only againſt the party himſelf, his lands and goods in his own hands, but in the hands of his heirs and terre-tenants, and againſt his executors and adminiſtrators, or if he had no executors or adminiſtrators, then againſt the poſſeſſors of his goods. Dyer 160. a. 11. Co. 93. a. Bunb. 322. But this muſt be underſtood, as to goods, where they were not aliened bona fide before the teſte of the execution. 8 Co. 171.

Proceſs ad computandum lay againſt terre-tenants.

And therefore where an officer and accomptant of the King died in arrear to him, the ſheriff having returned that there were no executors or adminiſtrators, proceſs ad computandum iſſued againſt the terre-tenants of his lands, although no judgment had been againſt himſelf in his life time. Dyer 324. And in Plowd. 321 a. where this caſe is cited, it is laid down ſtill more generally, as held therein, that if any perſon be accomptant to the King, or if any money, goods, or chattels perſonal of the King come to the hands of any ſubjeſt by matter of record or matter in deed, the lands of ſuch ſubjeſt are by the courſe of the Exchequer charged with the debt, and ſubjeſt to the King's ſeizure, in whoſever's hands they ſhall come after-wards, whether it be by deſcent, purchaſe, or otherwiſe; and that the law of the Exchequer is conſidered in ſuch caſe as the general law of the realm, and not as the law of the Exchequer only. But this ſeems to be laid down too largely, and is not at all warranted by the determination in Dyer *.

* See Favel's caſe in the time of Ed. III. Dyer 160. a. where it being found that, after he was appointed a collector, being languidus in extremis, he alien'd his lands, goods, &c. and died without heirs or executors, proceſs ad computandum was iſſued againſt the terre-tenants and the poſſeſſors of his goods. But it ſhould ſeem that this was on a ſuppoſition that the alienation was in order to defraud the King.

A ſcire

A *scire facias* iffued againft commiffioners of prize goods, grounded upon an inquifition, whereby they were found indebted to the King in a fum of money for prize goods, and to fhow caufe why the King fhould not have execution for this debt. And upon demurrer it was infifted that a *scire facias ad satisfaciendum*, which is a judicial writ, does not lie before the debt be determined upon record ; for that it is uncertain what the debt is, by reafon of the allowances that are to be made ; that * procefs of the pipe would not lie, which is not fo ftrong a procefs ; (for, by this courfe, body, goods, and lands might be taken into execution, when perhaps nothing was due ;) and that the auditing and ftating accounts is a judicial act, which ought to be done by the Barons, and not by inquifition. And by Hale, Chief Baron, a *diftringas ad computandum* is the ufual procefs. Hard. 228.

Scire facias ad computandum the property procefs where the King's debt is not determined.

All debts to the King on record bind the debtor's lands from the time they are contracted : for all lands being held mediately or immediately from the King, when any debt was recorded of any perfon it laid the eftate as liable to fuch debt as if it had been a refervation on the original patent or firft feudal donation. And therefore as the King

King's debts on record bind the debtor's lands. Gilb. Exc. 89. 122.

---

* Summons of the pipe iffued againft a man to levy £500 upon a *fuper fet* upon him by a colle:'or ; and a motion was made to fuperfede it, becaufe it could not be pleaded to, and it was fuperfeded : for, by Lord Chief Baron Hale, fummons of the pipe ought not to iffue but for a debt upon record, or a debt ftated and determined, and not for money due by matter *in pais*, as this cafe is ; wherefore if a collector in chief charge his under collector in account, or an accountant charge another together with himfelf for goods of the King's fold to him, and not paid for, fummons of the pipe fhall not iffue in thofe cafes, but a *fcire facias*, or a *diftringas ad computandum*, to which the party may plead ; for that thefe debts are not debts of record, but arife upon the accountant's charge only ; and fo here ; and a *fcire facias ad computandum* was awarded. Hard. 321. Same point determined, Hard. 504.

could

could feize the land for non-payment of the referved rent or fervice, fo he could feize it for any debt with which it was charged.

A truſt in fee extendible for the King's debt.

A truſt in fee of lands is liable to the King's debts by the courfe of the Exchequer; for the writ of *extendi facias* for levying the King's debts is of the debtor's lands, or of any land of which any other perfon was feized to his ufe ; tho' fuch an eftate does not efcheat in the cafe of felony, becaufe there is a tenant to anfwer the Lord's fervice. Hard. 495. 3 Chan. rep. 35. See 1 Vent. 132.

Or a term attendant on the inheritance.

Where a term is attendant on the inheritance, if the King extends the inheritance, he fhall have a right, to the term ; but if the term be mortgaged to one who has no notice of its being attendant on the inheritance, the mortgagee fhall hold it againſt the King. Prec. Cha. 125. 2 Vern. 389.

Where land granted by the King fhall not be extended for a debt due to him.

The King's receiver being indebted to him for arrears of his receipts, and being feized in fee of land, conveyed it in fee to I. S. who conveyed it to the King in fee ; and the receiver took it again from the King to him and his heirs; and afterwards the receiver became further indebted upon his account to the King. It was held in the court of wards that this land was not extendible for any of the faid debts; becaufe the land itfelf was never chargeable in itfelf, but in refpect of the perfon who was debtor, as in the cafe of a ſtatute ſtaple ; fo as, when the King took the lands, the debt was not thereby difcharged, but might be recovered againſt the debtor himfelf; but the land in the King's hands was not chargeable ; and then when the King conveyed the land over he could not againſt his own conveyance charge the land. But the Chief Baron doubted

it ;

it; and therefore the court decreed for the discharge of the land, without prejudice to the use of the Exchequer for the King's debt there. Hob. 45. But quere of this case, for it seems a strange determination.

A person being seized of lands made a conveyance with a power of revocation, and afterwards died indebted to the King, and without having revoked it; it was held that the land was extendible for the debt. 2 Roll. 295. Godb. 289.

Lands conveyed with power of revocation extendible.

But it was reckoned an abuse of the feudal prerogative if the King seized the lands or person of his debtor where there were goods sufficient to answer the debt: wherefore it was enacted by *magna charta*, c 8. and declared as part of the liberty of the subject, that the King or his bailiffs should not seize any land or rent for any debt, whilst the chattels of the debtor are sufficient to render the debt, and the debtor is ready to satisfy it.

King restrained by *magna charta* from seizing lands if goods sufficient.

By these means the abuse of the prerogative was totally hindered, and the King could not levy his debt on the land, whilst there were goods sufficient to answer it. From whence it became necessary to issue the summons of the * pipe against the debtor, which is a process against the goods only; and when any thing was *nihill'd* on the summons of the pipe, then, and not before, the second remembrancer's process, sometimes called the long writ, or prerogative process, issued, which is against body, goods, and lands, &c. heirs and executors. Gilb. Exc. 124.

Which introduced the summons of the pipe.

* But though the nature of the process of the pipe is so clearly pointed out, in this kingdom the summons of the pipe is against body, goods, and lands. When this practice commenced, which seems to have been originally through the mistake or ignorance of the officers, I have not been able to learn.

And

And thus the law stood in England until the 33 Hen. VIII. c. 39. by which it is enacted, that every suit for the King's debts, recognizances, obligations, or specialties, shall be made in the several offices and courts of his Exchequer, and other courts of revenue, under the seal of the said courts, by *capias*, *extendi facias*, *subpœna*, attachment, and proclamation, if need shall require, or any of them, or otherwise, as unto the said courts shall be thought by their discretion expedient for the speedy recovery of the King's debts.

From the time of making this statute of 33 Hen. VIII. c. 39. Lord Chief Baron Gilbert says the practice commenced of making *capias*, *fieri facias*, or extents, at the discretion of the court, to levy the King's debts.

Lord Coke says, that after this statute, the usual process to the sheriff was, " that you diligently, by the oaths of good and lawful men of your bailiwick, &c. inquire what goods and chattels, and of what price, he the said I. S. had in your bailiwick, &c. and you shall take them all into our hands, to the value of the debt aforesaid, and thereout cause to be made the debt aforesaid, &c. and if it shall happen that the goods and chattels of the aforesaid I. S. shall not be sufficient for the payment of the debt aforesaid, then you shall not omit by reason of any liberty, but you shall enter it, and by the oaths of good and lawful men diligently inquire what lands and tenements, and of what yearly value, the said I. S. had or was seized of in your bailiwick aforesaid, &c. and all and singular the aforesaid, in whose hands soever they shall be, you shall extend and take into our hands, &c. and you shall take the aforesaid I. S. so that you have his body to satisfy us of the debt aforesaid, &c."

But

But Lord Chief Baron Gilbert, with great reason, thinks this writ might have been used before the statute, without any violation of *magna charta*; because it seems so contrived that an inquisition should be found whether the debtor had any goods and chattels; and if upon the inquisition there were not sufficient found, then to extend the land, and take the body; and that therefore it seems to be a writ that was used in cases of necessity, before 33 Hen. VIII. but that since that statute there may be a *capias*, *levari*, or *extent*, without any inquisition touching the goods.

There are, according to L. C. J. Holt, five several forts of executions for the King. First a *capias ad satisfaciendum*, which commands the sheriff to take the body of the debtor. Secondly, a *fieri facias*, to fell his goods. Thirdly, an *extendi facias*, to extend his lands, &c. Fourthly, a writ, called the * long writ, comprising all the former. Fifthly, a *levari facias*, to levy the rents, issues, and profits of the lands, as in case of forfeiture of issues or of profits to be taken upon an outlawry; and upon this latter writ only, the cattle of a stranger *levant* and *couchant* upon the land may be taken. Comyn 51. 1 L. Raym. 306.

The writ of *extendi facias* or extent commands the sheriff to seize the lands and tenements, goods, chattels, and debts of the debtor, and to appraise them and extend and take them into the King's hands, until he shall be satisfied his debt; with a proviso that he fell no goods, until further process. It is said to be grounded on the aforesaid statute of 33 Hen. VIII. and is so mentioned to

* But note, this seems to be a different writ from the long writ, or treasurer's remembrancer's process.

be,

be in the end of the writ. It iffues from the equity fide
of the court, which is always open, and ufed formerly to
fit much longer than it does now. When the court does
not fit, they are made out, in England, upon the *fiat* of
a Baron, which is in the nature of an award of the court,
and they cannot be ante-dated before the *fiat*. 2 Strange
749. Bunb. 62.

Bonds to the
Crown when
introduced.
Gilb. Exc. 96,
105.

Before the time of Hen. VIII. there were few bonds
given to the Crown. But recognizances might be taken
to the Crown; for they were matters of record, and the
King could not take but by matter of record. But towards
that time, as the revenue increafed, and merchants were
obliged to make payments, the cuftomers and collectors
received bonds from the parties to the King. Thefe collectors were no more than bailiffs or receivers, and not as
juftices between the King and the party; and therefore
the acknowledgements before them were not in a court of
record; and there is this difference between them and
bonds of record, that thefe were immediately levied by
levari; but thofe not of record could not be levied by
levari, but a *fcire facias* was to iffue thereupon.

Difference
between
bonds to the
Crown and
recogniznn-
ces.
Gilb. Exc. 97.

And the reafon of the difference is, that where an obligation is acknowledged in a court of record, fuch recognizance is the fame as a judgment, the conufor being
perfonally prefent, and the court fuppofed to know him
as much as a defendant againft whom they give judgment.
And hence it is that the *levari* iffues, and all the other
prerogative procefs; and that the debt cannot be difcharged
until there be a receipt upon record. But where the
King's minifterial officer takes an obligation to the King,
fuch obligation is not of record; but when the officer
delivers fuch obligation into court, the time of fuch delivery

very is recorded; fo that if the obligation be juft and the conufor has nothing to fay againft it, nobody can controvert the time of its *lien*; becaufe the delivery is of record, and therefore it ought to bind from that time. But the obligation is no more than a warrant of attorney for the miniflerial or other perfon to delivery it of record; for being an act in *pais*, and not of record, the conufor may come in upon the return of the *fcire facias* and traverfe the obligation; but in this it differs from a warrant of attorney; for if a man forge a bond, and warrant of attorney, and then confefs judgment, the defendant can never deny the deed, if a *fcire facias* iffue after the year; but in this cafe there is no judgment upon the bond, for the bond is only delivered on record, and the judgment arifes only on the *fcire facias*.

When a bond or recognizance to the Crown is to be put in fuit, it is to be lodged in the Chief remembrancer's office, and from the time of delivery of the former (which is to be recorded by him) it binds the lands of the obligor. And thereupon *fcire facias* iffues thereon, directed to the proper county, or two to the fheriffs of the city of Dublin, let the place of abode of the party be where he will, to fummon him to fhow caufe, if any he can, why execution fhould not be.

Proceedings on bonds or recognizances to the Crown.

If the fheriff returns a *fcire feci*, then the officer is to enter rules to appear and plead thereon, as on an information; and the proceedings are as on fuch writs in fuits between party and party; with this difference, that when the three firft rules to plead are expired, a fourth rule is entered, viz. that the defendant plead in four days, or that judgment be entered againft him, without further motion;

On *fcire feci*-returned, rules to plead.

and if no plea be filed in thefe four days, judgment may be entered without procuring a certificate of no plea.

On two nihils returned, rule for judgment. If the fheriff returns two *nihils*, the officer will upon application enter a rule of courfe for judgment, if no plea in four days, which is the *quarto die poft*, for two *nihils* are deemed prefumptive notice, and equal to a *fcire feci*; and if no plea be filed in thefe four days, he will enter the judgment.

No judgment ufually entered. But upon thefe *fcire facias*, no judgment is ufually entered, which is lefs prejudicial to the debtor, for then he may obtain leave to plead; and it is as beneficial to the King to have an extent upon the bond or recognizance itfelf as upon a judgment.

Leave may be given to plead after judgment. If judgment fhould go upon two *nihils*, and the defendant makes affidavit that he has a reafonable and juft defence to make, and that he is ready to make it, the court will, upon motion, give him liberty to plead *.

Upon defendants pleading, procefs to ftay. *Gilb. Exc. 173.* If the defendant plead to the *fcire facias*, procefs is to ftay till the plea is determined; but if goods or lands be extended, procefs is not to ftay without fpecial order of the court, upon bringing into court the goods or the value

---

* In the cafe of the King againſt Thompſon, in this court, 26th November 1748, and 29 June 1750, the defendant having executed a bond to the King for performance of covenants to keep barracks in repair, writs of *fcire facias* iſſued thereon to the fheriffs of the city of Dublin in the uſual manner, and two *nihils* being returned, the King had judgment, and a writ of *levari* iſſued; but it appearing to the court upon a motion for the defendant, that the defendant lived at Waterford, and therefore was not fummoned, and the bond not being for the payment of money but for performance of covenants, which appeared by affidavits to have been performed, the court ordered the judgment and *levari* which iſſued thereon to be fet afide, and that the defendant fhould be at liberty to plead to the *fcire facias*. although the application was made upwards of four months after the *levari* had iſſued.

thereof,

thereof, or the mefne rents of the lands, or giving fecurity to abide the order of the court; and the reafon is, becaufe when a man pleads in difcharge of the *fcire facias*, he pleads in difcharge of the debt, and therefore the debt is in fufpenfe till the plea is determined; but where the lands and goods are extended, and he comes in to plead, it is to difcharge an execution executed; and therefore .nothing is to be ftay'd, until fecurity be given to anfwer the goods or the mefne profits of the lands.

If a bond be entered into to the Crown, with a warrant of attorney to confefs judgment, the warrant is brought to the officer, who enters a confent in his book of judgments, that judgment be forthwith entered up for the King, and that execution fhall iffue: In this cafe there is a *fcire facias* made out, figned by the officer and filed, but never fealed; which is in the nature of a declaration at common law, and the judgment is made up as thofe on the plea fide by *cognovit actionem*; becaufe they would not ftay the return of two *fcire facias*, to delay the King's execution, nor clog the rolls with two writs and two returns from the fheriff. *Proceedings upon bonds with warrant.*

Upon the writ of extent the fheriff is to hold an inquiry in order to find the lands, *&c.* and the yearly value thereof by examination of witnefles; which finding he is to return, and that he hath feized the lands into his hands for his majefty's ufe. *Proceedings upon an extent.*

And immediately upon his return, by the practice in England, a *levari* iffues, to levy the mefne rates half yearly, or oftener if it be required, until the principal debt, with cofts and damages, as the court fhall think fit, *And levari in England. Gilb. Exc. 170.*

Qq 2   be

be fatisfied; but the party may come and plead at the return of the extent, before any profits be actually levied.

How the
lands are put
in charge.
Gilb.exc.171.

And when the *levari* goes out, the remembrancer's office writes the lands in charge to the pipe, and from thence forward they are in charge on the fummons of the pipe, and the fheriff returns the iffues and profits of them annually; fo that it feems, that upon the firft iffues he anfwers to the *levari* before the Barons, and thofe iffues are drawn out into the pipe, in order to charge the fheriff, that the next year the fummons of the pipe may go out for the fame iffues, becaufe the lands are then within the complete charge of the fheriff.

Practice here
by granting
cuftodiams.

But the practice here is very different; for upon the return of the extent (which is here called a *levari*, and feems to be confounded with it) the folicitor for the Crown is to move for a *cuftodiam* and injunction; and the court will order that the clerk of the pipe do make out a *cuftodiam* of the lands, which is thereupon made to the collector of the diftrict, or of late more ufually to the folicitor, on account of the frequent changes of the collectors, in truft for the King, during the King's pleafure, at the yearly rent of * five fhillings; and the court will at the fame time order that the chief remembrancer do iffue an injunction, for putting the cuftodee or his affigns into the poffeffion of the lands. But for the further proceedings hereon, fee chap. 30. of *cuftodiams*.

* It is likewife ufual to infert in the *reddendum*, thefe words, " over and above the yearly rent and arrears of the premifes payable thereout to the King," but the infertion of thefe words feems to arife either from confounding the *cuftodiams* granted upon thefe *levaris*, with thofe granted upon feizures for arrears of the King's rents, or from a caution leaft the lands fhould be fubject to a crown or quit rent.

But

But in England, by the ftat. of 33 Hen. VIII. 39 a bond to the King is in the nature of a ftatute ftaple, and the Crown may iffue an * immediate extent upon it, at any time within a year after the bond was given. But if it be doubtful whether the condition be forfeited, or if it be profecuted after a year, a *fcire facias* iffues. But the crown may, even in fuch cafes, have an immediate extent, upon an affidavit made before a Baron, that the King's debt is in danger.

Bonds to the Crown in England in nature of ftatutes ftaple, and immediate extents go.

If goods be feized, and the extent returned, the court will, upon motion of the folicitor for the Crown, award a writ of † *venditioni exponas* to fell them. But Lord Chief Baron Gilbert fays, that in England, on return of the extent, a rule of fix days is to be given; and if the defendant do not appear at that time, then a *venditioni exponas* is iffued; but that if he appear and plead, a further rule of four days is given.

Venditioni exponas, when to iffue.

If two extents iffue againft a perfon bearing different *teftes*, and be delivered to the fheriff, and that which bears the lateft *tefte* be delivered firft to him, he fhould take an inquifition on that which bears the earlieft *tefte*, and make the common return upon it, viz. that he had feized the goods found into the hands of the King; and the fame goods being found by the fecond inquifition, to

If two extents iffue inquifition fhould be held on that of the firft tefte.

* It is faid that thofe immediate extents have been formerly, upon particular occafions, iffued in this kingdom; but as they are founded, as has been already mentioned, upon the 33 Hen. VIII. Eng. the legality of fuch proceeding here may be juftly queftioned.

† In England, terms for years are fold by *venditioni exponas*, upon extents for the King's debts. Bunb. 105. But fee Bunb. 71. where fuch writ was refufed.

return

return upon that, that they were feized upon the firft in-quifition; otherwife, if he return upon both inquifitions, that he has feized the goods, a *venditioni exponas* might iffue upon each, and he may be liable to be charged with both. Bunb. 323.

King's debt prior on record binds the lands of the debtor.
Gilb. Exc. 19.

As to the fecond particular, viz. Of the King's pre-cedence in executions. If the King's debt be prior on record, it binds the land of the debtor into whofe hands foever it come; becaufe it is in the nature of a feudal charge on the land itfelf, and therefore muft fubject every body that claims under it. But if the land were alien'd in the whole, or in part, as by granting a jointure, before the debt contracted, fuch alience claims prior to the charge, and therefore is not fubjected to it.

But if the fub-ject's debt on record be pri-or, the King's extent fhall be preferred, un-lefs it be after a *liberate*, ibid.

But if the fubject's debt be by judgment or recognizance, and prior to the King's debt, and the King extend the lands firft, the fubject fhall not by any after extent take them out of his hands: But if fuch judgment be extended, and poffeffion delivered to him by a *liberate*, he fhall hold it difcharged from the King's debt. But if the King's extent come before the poffeffion by *liberate*, the King's extent fhall be preferred, and the fubject wait till the King's debt be fatisfied.

The reafon of the difference. ibid.

The reafon of the difference is, becaufe the King's debt is in the nature of a feudal charge, which, if it come on the lands before the property of them is altered, feizes them as it might have done for the original fervice at firft impofed; but if there had been a lawful alienation before fuch debt, there it is not the feud of the tenant, and therefore fuch charge cannot affect it; therefore if there was a precedent judgment or recognizance and a *liberate* purfuant

purfuant, before the King's extent comes down, there it
cannot charge the lands, for the property is completely
altered by the extent of the fubject, which relates to the
time that the judgment was firft given, or recognizance
acknowledged, and is only an execution of it; but if the
King's extent had come before the *liberate*, he had charged
the land whilft it was in the hand of his debtor, and then
his charge would be fatisfied, as if it had been in the firft
feudal donation. And the lien upon lands by the fubject's
debt came in by the flat. of Weft. 1. for before that the
judgment did not bind the land; but the King's debt bound
the land before that ftatute, and the ftatute does not touch
the prerogative; and therefore the King has a power to
levy upon the lands, notwithftanding the preceding lien
by judgment; and therefore may feize lands that are
bound by a preceding judgment, whilft the lands are in
cuftody of the law on the *clegit*, and before the poffeffion is actually delivered to the creditor, as a fatisfaction
for his debt.

If A obtain judgment againft B, and B afterward enfeoff
C of his land, and then A affigns his judgment to the King,
in this cafe the King fhall extend but a moiety of the lands
in the hands of C. But if A had affigned the judgment to
the King, before B had enfeoffed C, the whole lands had
been liable; for the King by his prerogative could extend
all the lands of the debtor for his debt; but the feoffment
being made to C, before the affignment of the judgment to
the King, they were the lands of C before B became indebted to the King, and therefore the prerogative of the
King, which makes it a feudal charge, never affected
thofe lands, but they are fubject to the fame lien only to
which they were when it was only a debt due to A.
3 Leon. 239. 4 Leon. 10.

In what cafe the King fhall extend but a moiety. Gilb. Exc. 94.

A executed

Where the King's debt shall not be preferred.
A executed a bond to B, and C afterwards obtained two judgments of the fame term againft A, and B affigned his debt to the King; C took out two *elegits* upon his judgments, and extended both moieties of A's lands; and then procefs iffued out of the Exchequer for the debt affigned to the King; and the queftion was whether the King's debt fhould be preferred in this cafe, and it was determined that it fhould not. But the reafons upon which the determination was founded don't appear clearly from the reporter. One reafon affigned is, that the King's debt fhall be preferred when it is in equal degree, otherwife not. But this is a diftinction that does not feem to hold univerfally. Another reafon which feems a better one is, that the fubject's title was prior to the King's, and executed. But it was likewife faid that the 33 Hen. VIII. c. 39. which enacts " that the King's fuit fhall be preferred before any perfon's, and that he fhall have execution before any perfon; fo that his fuit be commenced before the other perfon's," abridged the prerogative in this refpect. Hard. 23, &c.

The King's execution binds goods from the *tefte*.
As to the King's execution of goods, it relates to the time of awarding thereof, which is the *tefte* of the writ, as it was in the cafe of a common perfon before the ftatute of 29 Car. II. c. 3. Eng. and 7 Will. III. c. 12. Irifh. For though by that ftatute no execution fhall bind the property of goods but from the time of the delivery of the writ to the fheriff. Yet, as this act does not extend to the King, an extent of a later * *tefte* fuperfedes an execution of the goods by a former writ.

* Bunb. 39, admitted, per curiam, that an extent binds from the *tefte*. 2 Strange 980. S. P.

And

And therefore, where an extent iſſued upon a ſtatute ſtaple at the ſuit of a ſubject, and the ſheriff ſeized the conuſor's goods, and after the day of the return, but before an actual return, and before a *liberate*, a prerogative writ iſſued againſt the conuſor for a debt due to the King, it was preferred. Dyer 67. b. 2. Ro. ab. 158.

Though after an extent at the ſuit of the ſubject, no - *liberate* being given.

But a ſale of a chattel, *bona fide*, by the King's debtor ſhall bind the King, ſo that his extent ſhall not reach it in the hands of the alienee. 8 Co. 171.

But not if alien'd *bona fide*.

A *fieri facias*, teſted 3d April, iſſued againſt a perſon, by virtue of which the ſheriff levied the goods, &c. but before a ſale, or the return of the writ, an extent came to the ſheriff at the ſuit of the Crown to levy the goods of the debtor, bearing *teſte* 2d May ; the ſheriff returned the ſpecial matter, on the *fieri facias*, and likewiſe upon the extent, in which it was ſaid that the debtor was poſſeſſed of the goods upon 30th April ; upon application to the court of Exchequer, he was obliged to amend his return, tho' there had been an inquiſition taken ; and in this caſe it was taken for granted that tho' the goods were levied by virtue of the *fieri facias*, three days before the *teſte* of the extent, yet that was no bar to the Crown. But the reporter adds a quære, if they had been ſold, becauſe then execution had been executed. Bunb. 8.

An extent will bind goods ſeized upon a *fieri facias*, but not ſold.

An extent having iſſued againſt a tenant, the landlord diſtrain'd for rent ; the next day the extent was executed, and the inquiſition found the goods then in the poſſeſſion of the tenant ; it was moved that the landlord might have the benefit of the ſtat. of 8 Ann. for his rent, notwith-

An extent will take place of a landlord's remedy under ſtat. of 8 Ann. Eng.

R r   ſtanding

ftanding this extent; but it was denied by the court. Bunb. 269.

Goods dif-
trained but
not fold liable
to the King's
extent, other-
wife of goods
pawned.

Upon a demurrer by the Attorney general to a plea to an inquifition upon an extent, the queftion was, whether the goods in the inquifition were legally feized into the King's hands, having been two days before diftrain'd for rent, but not fold. And it was determined by the court of Exchequer that they were, for that by the diftrefs they are in the cuftody of the law, and the property of them is not altered; and till an alteration of the property they are liable to an extent at the fuit of the Crown. But it was admitted that it would be otherwife of goods pawn'd before the *tefte* of the extent, becaufe the pawning is an alteration of the property. 2 Vefey 288.

Extent being
tefted on the
fame day as
an affignment
by commif-
fioners of
bankruptcy
fhall be pre-
ferred.

So where a man being indebted to the King by bond, an extent iffued againft him, and an inquifition found him poffeffed of goods; a third perfon pleaded that a commif-fion of bankruptcy had iffued againft him, that the goods were feized by virtue of a warrant from the commiffioners, and that the commiffioners had affigned to the affignees, on the day of the *tefte* of the extent; on demurrer, judgment was given for the King; becaufe the extent and affignment being on the fame day the extent is to be pre-ferred. Trem. P. C. 637. 2 Show. 481. Bunb. 33.

The goods of
a collector of
the land tax
feized by the
commiffioners
liable to the
King's debt
in preference
to the af-
fignees under
a commiffion
of bankruptcy
iffued againft
him.

The collector of the land tax being indebted to the commiffioners of the land tax, and having become a bank-rupt, they by virtue of a power given them by act of par-liament iffued their warrant, by which his goods were feized, and an affignment was in three days afterwards made of his effects by the commiffioners; and on an action of trover brought by the affignees, it was held by lord Hardwicke

Hardwicke and the court of King's-bench, that the collector was to be confidered as the King's fervant, and indebted to him, it being the King's money that he collected, and the allowance to him being made by the King; and that tho' the warrant was not to be confidered as equal to an extent, fo as to bind the goods from the date, yet that until an affignment the property was in the bankrupt, and the crowns hands were upon the goods, and created a lien before the affignment; and that the Crown was not bound by the acts relating to bankruptcy, not being named; and that upon this feizure all the right which the affignees had was to redeem the goods upon payment of the money, they being a pledge in the hands of the commiffioners for that purpofe.   2 Strange 978.

A deputy poft-mafter became indebted to the Crown, and an extent iffued againft him; he afterwards became a bankrupt, and the affignees under the commiffion obtained an order, that on payment of what was due upon his bond, the extent fhould be difcharged: upon motion to difcharge the order, it appeared by affidavit that the bankrupt had promifed alfo to difcharge a debt due from his father (who was alfo deputy poft-mafter and was fince dead) to the Crown, and for which a *diem claufit extremum* had iffued, and therefore it was held by the court that the affignees who ftood in his place ought not to have the benefit of this order, unlefs they would pay both debts purfuant to his promife.   Bunb. 337.

Affignees of bankruptcy not relieved againft an extent, but on paying a debt due to the Crown by the bankrupt's father, which he had promifed to pay.

It has been already obferved that if the King's debtor died, his debt was to be preferred in payment: But this muft be underftood, where the King's debt is on record; and therefore the King muft be firft fatisfied debts by judgments, ftatutes, recognizances, fines or amerciaments,

Of the preference of the King's debts by executors.

Rr 2   in

in his courts of record; and it would be a *devaflavit* in
the executor to pay other debts before them. But debts
due to the King, not of record, feem not neceffary to be
fatisfied before debts due to other perfons, where there is
no notice given of the King's debt; as where money is
due to the King for wood, tin, eftrays, &c. or for
amerciaments in a court baron or other court not of record.
Comyn 438.

The King
might take an
affignment of
a debt due to
his debtor.
Gilb. Exc.
167.

As to the third particular, viz. of the King's prerogative with regard to the debtors of his debtors, it is to be
obferved that the King, by an ancient prerogative, could
take from his debtors an affignment of any of their debts;
which was not allowed in the cafe of a common perfon;
becaufe it would have promoted maintenance; but it was
not prefumed that the King would maintain an unjuft
fuit.

Or extend
fuch debt in
aid of his debt.
Madox, 666.

And by another prerogative, if the King's debtor was
unable to fatisfy the King's debt out of his own chattels,
the King could betake himfelf to any third perfon, who
was indebted to his debtor, and recover of fuch third
perfon the debt due from him, in order to get fatisfaction
of the debt due to the Crown; and upon fuch recovery
had, fuch third perfon was acquitted againft the King's
debtor, and the King's debtor was acquitted againft the
King *de tanto*.

The King's
debtor might
have aid of
the Crown to
recover his
debts.
Madox 668.

Likewife, by the ancient ufage of the Exchequer, the
King's debtors or accomptants were wont to have writs
of aid, whereby to recover their debts of fuch perfons as
were indebted to them, in order to enable them to anfwer
the debts they owed to the King.

And

And the King may likewife have extents againft the debtor of the debtor of his debtor, and fo on as far as debts can be found; becaufe the fecond debtor, when his debt is feized, is a debtor to the King, fince the King can feize a *chofe in action*, and then the King may have an extent to feize the *chofe in action* due to fuch fecond debtor. But fee 4 Inft. 115.

King may have extents in aid ad infi-nitum.

But no obligation, recognizance, &c. for performance of covenants, though it be forfeited, or for any other matter than a debt due, can be affigned to the King by his debtor. 4 Inft. 115. 4 Leon. 9.

Bonds for performance of covenants.

And it is faid that thefe affignments of debts to the King are not favoured in law, when the King's immediate debtor is able to pay the debt; for by the affignment at the King's fuit, the body, lands and goods of the debtor to the King's debtor are liable to the King; whereas at the fubject's fuit, he could have had but his body, or goods, or half his lands. 4 Inft. 115. And fee 2 Leon. 31. 4 Leon. 80.

Affignments of debts to the King not favoured.

When any debt is found by † inquifition, and feized into the King's hands, either at the profecution of the King, or in aid of his debtor, procefs of *fcire facias* is to be awarded againft the party; which *fcire facias* fets forth the original debt, and then fets forth the inquifition taken of the debt due to the King's debtor, by virtue of fuch extent in aid.

Scire facias to iffue for fuch debts, and the form of it. Gilb. Exc. 177.

† Debts are not bound by the *tefte* of the extent, but by the caption of the inquifition. Bunb. 265.

Proceedings

Proceedings
of this kind
rare here.

Proceedings of this nature are very rare here, though frequent in the Exchequer in England, where feveral rules have been made concerning them, to prevent abufes, which would be proper guides to follow here; for which fee Gilb. Exc. 173.

Rules con-
cerning them
in England.

By thofe rules, he who affigns a debt is to take an oath that the debt affigned is a juft and true debt, and has not been formerly put in fuit in any other court; and that it is his own proper debt, originally due to him, *bona fide*, without any truft; and that he hath not received the fame nor any part thereof, except &c.

And by another, he who defires any debt to be found by inquifition in his aid, is to take an oath that he is juftly indebted to A, one of the farmers of the King's cuftoms, and that the fame is a juft and due debt, originally due unto the faid A, *bona fide*, without any manner of truft; and that B is juftly indebted to him originally and *bona fide*, without truft; and that the faid debt hath not been put in fuit in any other court; and that he hath not received the fame, and that C is much decay'd in his eftate, fo that unlefs a fpeedy courfe be taken againft the faid C the faid debt is in danger of being loft.

And by another, no further procefs is to be taken for debts in aid, than to inquire and feize the lands, debts, and perfonal eftate of him that is debtor to the King's debtor, or accomptant, unlefs it be by fpecial order made in open court.

The

The reafon of thofe rules, Lord Chief Baron Gilbert fays, was becaufe they made a ftate of the procefs; for many perfons indebted to the King affigned their own debts to the King, in order to get immediate extents againft their debtors; and therefore the court took care, that they fhould fwear them to be juft debts, and likewife that they fhould proceed no farther than an inquiry and feizure, and not to fell goods, without fpecial order of court.

And the reafon of them.

By another rule there, no debts without fpecialty fhall be affigned to the King; otherwife in the cafe of debts in aid.

Other rules.

By another, no debts without fpecialty fhall be found by inquifition for debts in aid, unlefs it be by order, upon motion in open court, and except it be for debts due to the King's farmers.

And by another, no immediate procefs of extent is to be awarded for debts in aid, but in cafes of extremity, and upon oath to be taken, as aforefaid.

The reafon of which is, becaufe no debts can appear to the court to be due to the King's debtor, without a fpecialty; and the prerogative of the King fhould not be abufed, by the affignment of fimple contract debts before they are tried; but where fuch debts are found to be upon extents in aid, there they are recorded upon the oath of a jury; but the court will not let fuch extent in aid iffue, before they are fatisfied that they are juft debts, and neceffary to be got in by the King's debtor.

And the reafon of them.

If

Sureties may have prerogative procefs againft principal.

If the principal debtor to the King fail, and his furetics pay the debt, they fhall have the prerogative procefs againft the principal. Comyn 390.

But as thefe immediate extents do not feem to have been practifed in thofe cafes in this kingdom, it will be fufficient to refer to Hard. 404. Bunb. 58, 127, 221, 225, 300. Comyn 388. where feveral points are determined concerning the iffuing them.

King's debtor obtaining a fraudulent extent in aid obliged to refund.

A farmer of excife having taken out an extent in aid againft a debtor of his that had failed, by which means the other creditors of his debtor were defrauded, upon a bill brought againft him in Chancery to be relieved, it appearing that he had fufficient eftate of his own to fatisfy the King's debt, he was decreed to refund. 1 Vern. 469.

But fuch matters are not examinable in the court of Chancery.

But in a later cafe, where one of the King's receivers took out an extent againft himfelf, and had a fimple contract debt due to himfelf found, and took out a *fcire facias* againft the executor of the fimple contract debtor; and the executor brought his bill in Chancery to be relieved, fuggefting that the proceedings were fraudulent, and on purpofe to gain a preference of creditors in a fuperior degree, and that the receiver was able to pay the King's debt himfelf; the receiver pleaded the proceedings in the Exchequer in bar to the relief prayed, but confeffed that he was able to pay the King's debt at the time of the extent; and the court allowed the plea. Prec. Cha. 47.

So

So where an extent in aid was taken out by a farmer of the hearth-money, against his own debtor, against whom a commiſſion of bankruptcy had before iſſued, but before the aſſignment of his effects; and the aſſignees brought their bill in Chancery to be relieved againſt the extents; the bill was difmiſſed, for that the court of Chancery had no juriſdiction in theſe cafes; and that any irregularities in the extents were properly examinable in the court of Exchequer, from whence they iſſued. Prec. Cha. 153. 2 Vern. 426.

C H A P. XXVII.

OF THE SEVERAL REMEDIES FOR THE RECOVERY OF THE KING's RENTS.

THE remedies for the recovery of the King's rents are either I. by diftrefs, II. by feizure, or III. by information.

The King may diftrain for his rent, on any lands of his tenant. And firft, as to the remedy by diftrefs; if the King has a rent fervice, rent charge, or rent feck, arifing out of lands, he may diftrain for it in all the other lands of his tenant, of whomfoever they be holden. 4 Inft. 119. 2 Inft. 132.

Whilft in the poffeffion of his tenant. But this is to be underftood, where the lands fo diftrained are in the actual poffeffion of the King's tenant, and not in the poffeffion of his tenant for life, or years, or at will, ibid.

Tho' the rent arife out of a franchife, &c. 1 Will. 307. And the King may referve a rent out of a franchife or matter incorporeal, as well as out of lands, and may diftrain on any other lands of the tenant for it.

The King may diftrain the goods of his under tenant off the land. 1 Ro. ab. 670. 2 R. ab. 159. If a man hold of the King by rent, and arrears incur, and the tenant make leafe to another, the King may diftrain the goods of the under-tenant for the arrears, in any place off the land holden.

By

By the ſtat. of 52 Hen. III. c. 15. Eng. it is declared
that it ſhall be lawful to no man to take diſtreſs out of
his fee, or in the high-way, nor in the common ſtreet,
but only to the King or his officers, having ſpecial authority to do ſo.

And by the ſtat. of 28 Ed. I. c. 12. Eng. it is further
provided, that all diſtreſſes which are to be taken for the
King's debt ſhall not be made by beaſts of the plough,
ſo long as a man may find other; and that too great diſtreſs ſhall not be taken for his debts, nor driven too far;
and that if the debtor can find ſurety, until a day before
the day limited to the ſheriff, the diſtreſs ſhall be releaſed
in the mean time.

Diſtreſs for the King's debt not to be made by beaſts of the plough, nor too great diſtreſs taken.

It is obſerved by Mr. Barrington in his obſervations
upon the ſtatute *de ſcaccario*, that at this time the ſheriffs
generally farmed the King's revenue, and conſequently
were guilty of thoſe enormities and exactions which the
farmers of the publick revenue have in all countries been
juſtly charged with.

Great abuſes committed by ſheriffs, who farmed the revenue.

Replevin does not lie againſt the King, nor where the
King is party, nor where the taking is in right of the
King, and yet it is lawful in ſuch caſe for the ſheriff *primâ
facie* to grant replevin, but when it is ſhewn that the King
is party, or that the taking is in right of the King, there
the ſheriff ſhall ceaſe. Br. repl. p. 33.

Replevin lies not againſt the King.

And Lord Chief Baron Gilbert aſſigns this reaſon; for
that the diſtreſs for the King's debt is in the nature of a
levari, which is a writ of execution; and that conſequently

The reaſon why.

Sſ 2   no

no replevin lies againſt the King, any more than it does for goods taken in execution at the ſuit of common perſons.

Attachment will be granted againſt the party ſo re-plevying.

And in the caſe of a conſtable, who being fined by the commiſſioners of the land tax, and diſtrained, afterwards replevied, it was held by the court of Exchequer in England, that if there be a diſtreſs for any duty to the Crown, the perſon diſtrained cannot replevy; and that if he does, an attachment ſhall be granted againſt him for the contempt. Bunb. 14.

Or againſt the ſheriff.

And in the caſe of the King againſt ſir Thomas Denny in this court, 28 May 1756, a diſtreſs being taken for an arrear of quit rent, alleged to be due to the Crown, by virtue of a warrant from the collector of the diſtrict, was replevied by the ſheriff of the county, whereupon an attachment was awarded againſt the ſheriff; although it was infiſted that no rent was due at the time of the diſtreſs.

And in the caſe of the King againſt the ſheriff of the county of Roſcommon, &c. Eaſter term 1757, on a like application, an attachment was awarded, unleſs cauſe; but no further proceedings were had, the ſheriff having made a proper ſubmiſſion.

The proper remedy for the party in ſuch caſe.

But if the party, whoſe goods are diſtrained for the King's debt, have any matter of relief to ſhow, the uſual method is to depoſite the ſum for which the diſtreſs is taken, and then to apply to the court by counſel, on affidavits of the facts; and if it appears to the court that the diſtreſs was taken improperly or unjuſtly, they will relieve the party, by ordering the ſheriff or other officer of the Crown, by whom the diſtreſs was taken, to return the

the money, or by attaching him, according to the circumftances of the cafe. See the cafe of the King againft
Lord Ikerrin Trin. 1746.

Secondly, as to the remedy by writ of feizure; hereto- Writs of fei-
fore where lands, out of which a rent was payable to the zure.
Crown, were returned as wafte, the Crown was under
great difficulty in the recovery of the rent; which being
reprefented to government, on the 21ft of February 1661,
the following order was made, viz.

Whereas feveral lands in Ireland, liable to pay his Rule of 21
Majefty's new quit rents, are by the commiffioners ap- Feb. 1661, as
pointed for affeffing the faid lands returned wafte, to the where lands
leffening that branch of his Majefty's revenue; which are returned
being made known to the lords juftices and council, their wafte.
lordfhips by their order of the † 22 February inft. did require and direct the Lord Chief Baron and the reft of the
Barons of this court, to caufe all fuch lands to be feized
into his Majefty's hands, and difpofed of for his advantage
and benefit, until the quit rent with the arrears thereof
be difcharged, as by the faid order remaining of record in
the Lord Treafurer's remembrancer's office of this court
appeareth; in purfuance whereof it is ordered by the
court, that the faid remembrancer do forthwith iffue *
feizures to the fheriffs of the refpective counties of this
                                          kingdom

† So in the original, but the date muft be a miftake.

* It does not feem very clear, in what light thefe feizures are to be confidered,
or upon what legal authority they are founded; for an order of government folely
cannot be deemed fuch. If they are to be confidered as a prerogative procefs, in
the nature of an immediate extent or levari, for the levying the debt due to the
Crown, they feem not warranted by the laws of this kingdom; fuch procefs being,
as has been fhown in the laft chapter, grounded on the ftat. of 33 Hen. 8. 39 Eng.
and even in England, it feems to be held that a levari ought not to iffue for a fee
                                                                            farm

kingdom wherein such lands do lie, to seize the same into
his Majesty's hands, returnable the next Easter term; and
in order thereunto his Majesty's Auditor general is to return to the said remembrancer a full and perfect list of the
lands so returned waste, to the end that his Majesty's revenue may the speedier be * performed.

And the constant usage hath been since, if a sufficient
distress be not to be had, upon a proper affidavit of the
fact, and upon a *constat* from the Auditor general of the
charge upon the lands, and the arrear due, and upon
counsel's motion thereon, for the Court to award a writ of
seizure, without a *scire facias*, directed to the sheriff, to
seize the lands.

Seizure of incorporeal thing.   So likewise if the arrear be out of the rent of a rectory,
tithes, fair, ferry, fishery, or other incorporeal thing, of
which there can be no distress, upon such *constat* and
motion the court will grant a writ of seizure.

Constat of the rent in lieu of an office.   And the *constat* is deemed a sufficient finding of the
King's title upon record to supply the want of an office,
which in such cases has been held necessary to entitle the
King.

Seizure where distress is rescued.   It has been doubted whether upon a rescue of a distress
taken for the King's rent the court could grant a writ of
seizure; but in Trin. term, 18 July, 1750, in the case of

farm rent; but that a distress is the proper remedy. Bunb. 348. If they are considered
as a re-entry by the Crown, for a breach of condition, in non-payment of the rent, it
may be questioned whether there should not be a previous inquisition finding the rent
to be due. See 2 Inst. 205. Cr. El. 220, 855. Cr. Car. 100. Poph. 53. 5 Co.
56. b. 2. Ro. ab. 184, 215.

* So in the original.

the

the King againſt the lands of Clonbeg, it was determined that in ſuch caſe a writ of ſeizure is a proper remedy to recover the arrear ; for that where the diſtreſs is reſcued the lands are in effect as waſte ; and it was likewiſe held that it is not neceſſary in ſuch caſe to give notice to the tenants before the iſſuing the writ ; for that the diſtreſs and reſcue are a ſufficient notice that the rent was due ; and that if a conditional order were made, the punctuality and nicety required in ſerving it, and the opportunity given thereby to the parties to conteſt the ſervice and the charge, would put the Crown to more expenſe than the value of the arrear ; ſo that the King would loſe by recovering his rents.

And the court will alſo in ſuch caſe grant attachments againſt the perſons committing the reſcue ; as was done 3d Feb. 1756, in the caſe of the King againſt the lands of Bonane in the county of Sligo.

And attachment.

The affidavit of the diſtreſs and reſcue is generally made by the collector's driver, who is to ſwear that he was authorized and empowered by the collector to diſtrain the land ; and this affidavit is to be filed in the ſecond remembrancer's office.

Upon affidavit of the driver.

Formerly it was uſual for the collectors of diſtricts, by order from the commiſſioners of the revenue only, and without a writ of ſeizure, or any writ or proceſs from the court, to ſeize rectories, tithes, &c. for arrears of Crown rent, and to ſet them to tenants. But by a rule made in this court, 4 Dec. 1711, it was declared that ſuch practice was contrary to law, and the ancient and conſtant uſage and courſe of the Exchequer : for that when any rectory, vicarage, or lands, are in arrear to the Crown, the commiſſioners

Rule of 4 Dec. 1711, reſtraining the commiſſioners from ſeizing, &c.

fioners cannot feize them, but muft apply to the court of
Exchequer in the ufual manner, who thereon iffue their
feizure, directed to the fheriff of the county where the
lands lie ; on return of which the court grants a *cuftodiam*
thereof to fome perfon in truft for the Crown until the
arrear be paid.   And the court declared that the commiffioners ought for the future to proceed on all fuch occafions according to the ancient ufage of the Exchequer.

Attachment to the fheriff, and writ of affiftance to aid the purfuivant.   In fome fpecial cafes the court have directed the attachment to iffue to the fheriff; as was done in two cafes
21 Nov. 1688.   And they will upon fpecial application
order a writ of affiftance to iffue to aid the purfuivant.

Proceedings on the feizure by inquifition, *cuftodiam*, and injunction.   Upon the writ of feizure, as upon a *levari facias*, the
fheriff is to call a jury, and hold an inquiry, in order to
find the lands, &c. and their value ; which finding he is to
return into this court, and alfo that he has feized the lands,
&c. into his hands for his Majefty's ufe ; upon which
return and finding a *cuftodiam* is to be made out, and an
injunction iffued, as is mentioned in the laft chapter.

Whether the tenants in poffeffion may be re-moved by the injunction.   And it is faid that on this injunction the fheriff may
remove and turn out every tenant on the lands, although
they fhould have legal interefts fubfifting prior to the
feizure ; for that the King's feizure is in this cafe in the
nature of a re-entry for the non-payment of the rent ; and
that the King comes in then by title paramount to all
leafes made by his tenant.   However this power is rarely
put in execution, and only in defperate cafes ; the ufual
method being to get an order for the tenants to pay their
rents or fet out their tithes as the cafe is.

                                                            And

And when the cuſtodee of the Crown is put into poſſeſſion of the lands, &c. he may ſet them to tenants during
his intereſt therein, without any order or further power
from the court whatſoever; and the uſual method is by
publick cant, after publick notice given, from year to
year. But if they be ſet at an under value, or there be
any improper methods uſed in ſetting them, the court
upon proper application will relieve the party injured.

Cuſtodee may ſet the lands, without further order.

Thirdly, the King may have remedy by information in
debt, in the name of his Attorney general, againſt the
grantee or alienee of the land, or againſt his heir, or
againſt an incumbent of a rectory, &c. for rent accrued
during their time; as he may likewiſe againſt the Executor, &c. of the grantee, for arrears due in the time of
the teſtator; or againſt the heir, if the Executor, &c. has
not aſſets ſufficient; and upon judgment obtained in ſuch
information, the Crown may have a *levari*.

Of the re-medy by in-formation in debt.

Or an Engliſh information may be brought in the equity
ſide of the Exchequer, againſt the heir, executor, or ad-
miniſtrator of the grantee or alienee, for rent incurred in
the time of the anceſtor; which is the more eligible
method, becauſe the Crown may thereby have a diſcovery
of aſſets *.

Or Engliſh bill.

---

* Mr. Howard having, in the exerciſe of the duty of his office, had frequent
occaſions to take the opinions of the moſt eminent council, as well in England as
this kingdom, with regard to the proper methods of proceeding for the recovery of
the King's rents, under different circumſtances, and thoſe opinions having received
the ſanction from time to time of judicial determinations by this court, he apprehends it may not be uſeleſs to inſert here an abſtract of them.

And it was holden, that the cattle, &c. on the lands out of which the rent is due, in the hands of the heir of the grantee, may be diftrained for rent due, as well in the time of his anceftor as in his own time ; and that the cattle, &c. on other lands of the heir in his poffeffion may be diftrained for rent incurred in his own time. And that the fame holds as to the alienee and his heir.

That where a diftrefs is to be taken on other lands of the King's tenant, than thofe which are charged with the rent by the grant, and in another diftrict, in fuch cafe, as the collectors are confined to their feveral diftricts, fo that the collector of that where the lands charged lie cannot diftrain the other lands, nor the collector of that where the latter lie cannot diftrain for rent not given in charge to him, it is proper that the commiffioners of the revenue do give a fpecial authority for the purpofe to the collector of the diftrict where the diftrefs is to be made.

That where the King grants a rectory or tithes to a bifhop and his fucceffors, in truft for the incumbents having cure of fouls in his diocefe, the King may diftrain the cattle of the incumbent in any lands in his actual poffeffion (though no part of the glebe land, or any land belonging to the rectory) for any rent which became due during his incumbency, but not for any arrear which accrued in the time of his predeceffor ; but that the glebe lands may be diftrained for the arrears due in the time of his predeceffor.

That where a grant is made by the Crown of impropriate or appropriate tithes, under the act of fettlement, to a bifhop and his fucceffors, to the ufe of the incumbent and his fucceffors, " the bifhop, &c. to permit the incumbents to take the tithes, &c the incumbents, &c. from time to time indemnifying the bifhops," by which words it fhould feem to be the intention of the grant, that the bifhop fhould be liable, yet as the act of fettlement (ante 207) vefts the tithes in the incumbents, and the rents are referved from and payable by the incumbents, who fhould therefore be confidered in conftruction of law as having thofe tithes, and as faling under the defcription of the *levari*, which mentions fuch lands, &c. as the debtor had, · &c. (Godb. 294), and as there is a clear remedy againft the incumbent, no attempt fhould be made to recover the rent from the bifhop.

In the year 1762 feveral large arrears of Crown rent for fifty years and upwards having grown due out of feveral rectories and tithes, and efpecially in the county of Dublin, of which there did not appear any grants from the Crown, nor any evidence of the title of the Crown thereto, except the ancient rent-rolls, and the act of fettlement, by which all forfeited rectories and tithes are vefted in the incumbents having cure of fouls ; in this cafe it was agreed by thofe opinions :

1. That if the rectory appeared to be impropriate or appropriate fuch rent-rolls will be good evidence of the Crown rent anciently referved.

2. That

2. That thefe rent-rolls, confidering the length of time fince any rent was paid to the Crown, would not be fufficient evidence in themfelves to fupport the title of the Crown, efpecially as there are other circumftances the prooi of which, or fome of them, might be expected, as that the rectory was impropriate before 1641, that it was feized and fequeftered during the time of that rebellion, the attainder of the proprietor, &c.

END of the FIRST VOLUME.

www.ingramcontent.com/pod-product-compliance
Lightning Source LLC
Chambersburg PA
CBHW030857270326
41929CB00008B/460